📞 01603 773114   cninformationstore
email: tis@ccn.ac.uk   @CCN_Library   ccnlibraryblog.wordpress.com

21 DAY LOAN ITEM

Please return <u>on or before</u> the last date stamped above

A fine will be charged for overdue items

Thieme
Stuttgart · New York

Library of Congress Cataloging-in-Publication Data is available from the publisher

This book is an authorized and revised translation of the German edition published and copyrighted 2002 by TRIAS Verlag, Stuttgart, Germany. Title of the German edition: Erfolgreiche Wundheilung durch Maden-Therapie. Biochirurgie: Die wieder entdeckte Behandlungsmethode bei diabetischem Fuß und anderen schlecht heilenden Wunden.

Translator: Suyzon O'Neal Wandrey, Berlin, Germany

© 2004 Georg Thieme Verlag,
Rüdigerstrasse 14, 70469 Stuttgart, Germany
http://www.thieme.de
Thieme New York, 333 Seventh Avenue, New York, NY 10001 USA
http://www.thieme.com

Cover design: Matina Berge, Erbach
Typesetting by Satzpunkt Ewert GmbH
Printed in Germany by Druckhaus Beltz, Hemsbach
ISBN 3-13-136811-X (GTV)
ISBN 1-58890-232-3 (TNY)   1 2 3 4 5

Important note: Medicine is an ever-changing science undergoing continual development. Research and clinical experience are continually expanding our knowledge, in particular our knowledge of proper treatment and drug therapy. Insofar as this book mentions any dosage or application, readers may rest assured that the authors, editors, and publishers have made every effort to ensure that such references are in accordance with **the state of knowledge at the time of production of the book.**
Nevertheless, this does not involve, imply, or express any guarantee or responsibility on the part of the publishers in respect to any dosage instructions and forms of applications stated in the book. **Every user is requested to examine carefully** the manufacturers' leaflets accompanying each drug and to check, if necessary, in consultation with a physician or specialist, whether the dosage schedules mentioned therein or the contraindications stated by the manufacturers differ from the statements made in the present book. Such examination is particularly important with drugs that are either rarely used or have been newly released on the market. Every dosage schedule or every form of application used is entirely at the user's own risk and responsibility. The authors and publishers request every user to report to the publishers any discrepancies or inaccuracies noticed.
Some of the product names, patents, and registered designs referred to in this book are in fact registered trademarks or proprietary names even though specific reference to this fact is not always made in the text. Therefore, the appearance of a name without designation as proprietary is not to be construed as a representation by the publisher that it is in the public domain.
This book, including all parts thereof, is legally protected by copyright. Any use, exploitation, or commercialization outside the narrow limits set by copyright legislation, without the publisher's consent, is illegal and liable to prosecution. This applies in particular to photostat reproduction, copying, mimeographing, preparation of microfilms, and electronic data processing and storage.

Preface

This volume and the German edition that preceded it are landmarks in maggot therapy history. Together, they represent a milestone in the 21st century revival of maggot therapy. Seventy years ago, no one ever imagined that there would be the volume of information or the popular demand to warrant an entire book on the topic of maggot therapy. Just 10 years ago, almost no one believed that the world would again see the widespread use of maggot therapy or the need for an entire book on this topic. But this book is now surely warranted, given our expanding knowledge of this awesome medical practice, and the growing interest and use of biodebridement to aid in the care and healing of wounds unresponsive to more conventional medical and surgical therapies.

Drs. Fleischmann and Grassberger brought their wealth of knowledge and experience to the original German edition, which explored the history, biology, pathology, mechanics—the science, philosophy, and art—of maggot therapy. Now, 2 years later, this volume expands on that work with the results of new research, more chapters, and additional references, and brings the entire body of work to the English-speaking world. Case histories illustrate the varied uses of maggot therapy. Technical problems and their solutions are discussed.

The clinical photographs are not for the weak-hearted, nor for those with weak stomachs. They are included because they accurately illustrate the wounds that often require maggot therapy: gangrenous, life- or limb-threatening wounds that have progressed for weeks, months, or even years despite the best of modern medicine and surgery. No matter how much the reader may be disgusted by maggots, after reading the case reports the reader will come to agree with maggot therapy patients and practitioners: the smelly, draining, gangrenous wounds are much more disgusting than the maggots.

Preface

Whether a student of biology, a prospective patient, or a medical professional considering the use of maggot therapy in your own practice, this book will inform, amaze, and entertain you. You will be engrossed. Enjoy it.

Autumn 2003 *Ronald Sherman*

Contents

1 Introduction 1
Nature's Vast Pharmacy 1

2 Maggots 3
On Flies and Maggots 3
Development Cycle of the Fly 4
Transmission of Disease by Flies 7
Phaenicia (Lucilia) Sericata (Green Blowfly) 8
Larvae of Phaenicia Sericata 9

3 Maggots and Wound Healing 14
History of Maggot Therapy 14
Stimulation of Wound Healing 25
Medicinal Uses 27
Production of Sterile Maggots 28
Myiasis 29

4 Clinical Application of Maggots 32
The Classical Free-Range Maggot Dressing System 32
The Containment-Bag Maggot Dressing System (Biobag) 34
Indications for Maggot Debridement Therapy 37
Adverse Effects and Risks of Maggot Therapy 63
Perspectives 64
Cost Effectiveness 65

Contents

5 Appendix 66
Frequently Asked Questions 66
Suggested Reading 71
Sterile Maggot Suppliers 75
Glossary 77
Index.. 82

1 Introduction

Nature's Vast Pharmacy

Many people have their doubts about using maggots as medicine. However, they should bear in mind that a vast number of drugs come from nature's pharmacy.

Hirudin, a pharmaceutical agent used to dissolve blood clots, is a good example. This naturally occurring anticoagulant was isolated from the saliva of the medicinal leech. Hirudin serves to keep the blood flowing freely so that the leech can easily ingest it after biting its host. The salivary secretions of bats and snakes contain similar anticoagulants. To this day, the well-known antibiotic penicillin is fermented from a mold that produces the compound to kill its bacterial competitors. The *Cantharanthus roseus* plant is a natural source of antineoplastic alkaloids (vinblastine and vincristine) used to destroy malignant tumors. The drugs derived from this plant alone net more than $ 180 million in sales each year. The list goes on and on.

Few people realize just how much the pharmaceutical industry depends on natural organisms for drug manufacture. Of all prescription drugs sold, 25 % are derived from plants, 13 % from micro-organisms, and 3 % from animals. Accordingly, over 40 % of our pharmaceutical drugs come from nature.

Insects are a real treasure-trove of raw materials for drug manufacture. They produce a variety of active biochemicals, including sex pheromones, alarm pheromones, defensive substances, and venoms. For example, honey bee venom has long been used to treat arthritis, and butterfly, beetle, and wasp venom extracts appear to be effective in fighting cancer.

How and why did these useful substances come to exist? The secret lies in organic evolution. In the course of its phylogenic history, each living organism has evolved into a living chemical factory that produces the substances it specifically requires to survive in a hostile environment.

Millions of years of natural selection and adaptation have turned the most diverse organisms into chemists of immeasurable ingenuity—true masters in solving some of the same biological problems that also undermine the health of humans and other organisms.

The world wars waged in the first half of the 20th century brought devastation and great suffering to humankind. War injuries often resulted in incurable infections of the bone. In many cases, limb amputation was the only recourse for their survival. This dark picture was slightly brightened by countless reports of soldiers whose maggot-laden wounds were free of infection. Soon it became clear that the maggots were responsible for saving many lives and limbs. Thus maggot debridement therapy (MDT) was born.

Indeed, Hippocrates' maxim is as applicable today as it was some 2 400 years ago: "*medicus curat, natura sanat*" (the doctor administers the cure, nature does the healing).

2 Maggots

On Flies and Maggots

The Greek philosopher Aristotle (384–322 BC) named the fly "*Diptera*". The Greek word "*dipteron*" means "two-winged", referring to the single pair of functional wings that distinguish the fly from virtually all other insects. When working on his system of taxonomic classification of living organisms, or *Systema naturae*, published in the 18th century, Carl von Linné (*Linnaeus*, 1707–1778) adopted *Diptera* as the taxonomic name for the order of insects to which all true flies belong.

The origin of the dipterans is unknown. The oldest known fossils date back to the Triassic period and are some 210 to 220 million years old. These relicts mainly consist of the wings of adult flies. Signs of other stages of early fly development are practically non-existent.

The coexistence of flies with humans and domestic animals (synanthropy) has left notable marks in the history of humankind. Flies gained a reputation as pests, parasites, and carriers of harmful diseases. Written records and cult objects surviving from various periods testify to the explosive multiplication of fly populations during wars, famines, and other catastrophes. In all of these periods, people's attention was most strongly drawn to the seemingly apocalyptic plagues of flies that occurred throughout history.

The historical narratives cited below underline the timelessness and the global impact of the fly problem. The best known reference to plagues of flies is probably that in the Old Testament book of Exodus:

> "This is what the Lord says: Let my people go so that they may worship me. If you do not let my people go, I will send swarms of

flies on you and your officials, on your people and into the houses. The houses of the Egyptians will be full of flies, even the ground where they are. And the Lord did this. Dense swarms of flies poured into Pharaoh's palace and into the houses of his officials, and throughout Egypt the land was ruined by the flies." Exodus 8:20–21 and 8:24, NIV Version, 1984

Fly populations multiply rapidly in warm weather and on corpses. The military physician Ambroise Paré (1510–1590), who reported on the Battle of Saint Quentin (1557), described this phenomenon as follows:

"We saw more than half a league round us the earth all covered with the dead; and hardly stopped there, because of the stench of the dead men and their horses; and so many blue and green flies rose from them, bred of the moisture of the bodies and the heat of the sun, that when they were up in the air they hid the sun. It was wonderful to hear them buzzing; and where they settled, there they infected the air, and brought the plague with them."
(Quoted from: Ambroise Paré, Journeys in Diverse Places. The Harvard Classics. 1909–14)

Development Cycle of the Fly

There are over 100 000 species of flies, representing a variety of shapes and sizes (morphology), of habitats, and of behaviors. Yet they all have in common a 4-stage "complete" (holometabolous) metamorphosis, by which they develop through the stages of egg, larva, pupa, and finally adult (Fig. 1). As an example, the blowfly life cycle will be described in more detail. Female flies lay masses of up to 200 eggs, usually on dead bodies and decaying meat, but also on open wounds. Flies have special sensory organs that enable them to immediately recognize decayed flesh that is suitable for feeding and egg laying. The adult female unfurls its ovipositor (Fig. 2) and lays ("blows") hundreds of its eggs on the meat. Hence the name *blowfly*.

Development Cycle of the Fly

Fig. 1 Immature fly stages. A: Eggs of *Phaenicia (= Lucilia) sericata*. B: Hatching first instar larvae

A female fly can lay up to 3 000 eggs in her lifetime. The number of eggs laid is determined by the size of the female and by the quality and quantity of food she consumes. On a protein-rich diet, a female fly may start to lay eggs as soon as five days after

2 Maggots

Fig. 2 Adult female *Phaenicia* (= *Lucilia*) *sericata* with projecting ovipositor.

emerging from the pupal case. The time required for egg and larval development is mainly determined by ecological factors, such as environmental temperature and humidity. Fly eggs generally hatch into maggots in 12 to 24 hours, the maggots mature to pupae approximately one week later. Normally pupae transform into adult flies within one to three weeks; but under unfavorable conditions, it can take weeks or even months for this process to occur.

The adult fly is whitish-gray in color when it initially emerges from the pupal case. The cuticle (external chitinous shell) then stretches, hardens, and dries, resulting in the typical metallic appearance of the adult blowfly. The life span of a fly is roughly one to two months. Accordingly, four to eight generations of flies can develop during the major breeding months of May through October.

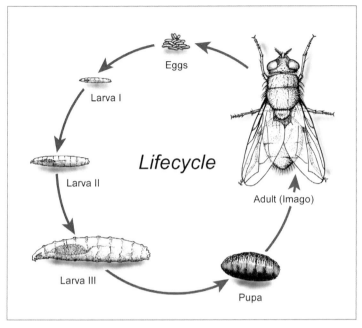

Fig. 3 *Blowfly lifecycle*

Transmission of Disease by Flies

Flies have a great impact on human and animal health. Most of the known fly species are harmless to humans, but around 11 000 species can cause disease in one or more ways: 1) as vectors, carrying (and sometimes breeding) parasites within their own body, which they then inject into their host as they suck mammalian blood; 2) as fomites, mechanically transporting infectious bacteria or viruses from one site to another as they visit dumpsters, excrement, and prepared food; or 3) as parasitic larvae (myiasis), living and feeding on the tissues of live vertebrate hosts. The bloodsucking flies are responsible for the transmission of a number of harmful diseases, such as malaria, filariasis, onchocerciasis, leishmaniasis, and African trypanosomiasis. They transmit microbial pathogens or parasites that enter the host skin or circulation when the flies bite.

Phaenicia (Lucilia) sericata (Green Blowfly)

Blowflies (Calliphoridae) "blow" (lay eggs) on rotting organic material, especially animal tissue. The female fly has a keen sense of smell that helps her find a suitable host, such as an infected wound or decaying meat, located even very long distances away.

Members of the genera *Lucilia* and *Phaenicia* are commonly known as green blowflies (greenbottle flies) because of their iridescent golden-green color. The eyes of the female fly are wide-spaced and are separated by the forehead (Fig. **4**), whereas those

Fig. **4** *Phaenicia sericata* (green blowfly, greenbottle).

of the male are close-set. There are many different species within these two genera, each with their own distributions and habits. There may even be multiple strains or subspecies within each species, although it has not been possible to identify strain differences morphologically. *Lucilia* and *Phaenicia* are synanthropes, living in close association with humans.

The earliest blowflies (Calliphoridae) probably fed exclusively on decaying flesh from the bodies of dead vertebrates, as described hundreds of years ago in one of the first European medical textbooks, *Hortus Sanitatis* (Mainz, Germany, 1491).

Larvae of *Phaenicia sericata*

The larvae of *Phaenicia sericata* (classified by some as *Lucilia sericata*) have a typical maggot shape, i. e., they have a sleek, tapering front end (head) and a blunt, flattened back end (tail). The body of the maggot consists of 12 segments without any clear division between the head and the other body segments. A furrow divides the head into a left and a right lobe; the mouth is situated inferiorly, at the base of the furrow. The complex cephalopharyngeal skeleton, the mouth hooks of which are visible externally, is operated by a strong muscular apparatus. The cephalopharyngeal skeleton helps the maggots move about. Annular spicules on each segment of the body keep the maggot from sliding backward.

Maggots breathe through apertures called spiracles, which are located at the anterior and posterior ends of their body. The posterior respiratory spiracles of growing maggots are often mistaken for eyes.

The head of the blowfly maggot contains primitive sensory organs that only allow the maggot to distinguish between light and darkness. Unlike the adult fly, maggots always move away from light (negative phototaxis). Several maggots unite to form feed-

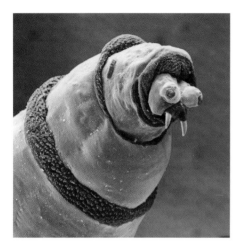

Fig. 5 Front end of a blowfly maggot as seen under a scanning electron microscope.

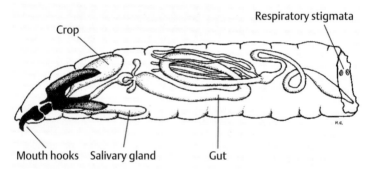

Fig. 6 Maggot anatomy

ing communities. The maggots feed by dipping their front end into the liquid nutritive substrate while breathing through their posterior respiratory apertures.

The fact that maggots require air to breathe should always be borne in mind when dressing wounds with live maggots.

Digestive enzymes are continuously produced by two labial glands (salivary glands) and secreted into the surroundings. A powerful pharyngeal pump sucks in the liquefied, bacteria-laden food, which is then passed through a filtering system that concentrates the nutrients roughly five-fold. This feeding strategy allows the blowfly maggot to ingest a quantity of food equivalent to half its body weight within five minutes.

The larvae of *P. sericata* have a prominent crop that can hold and store large a quantity of food until it is later needed. The glands that produce the digestive enzymes, the powerful suction apparatus, and the extremely distensible crop are evolutionary adaptations that reflect the main purpose of the larval stage: name-

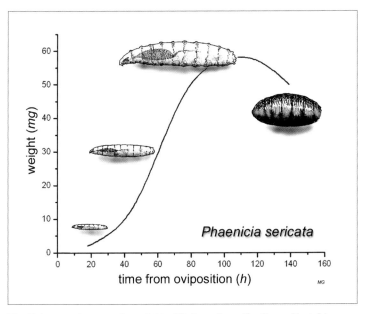

Fig. 7 Average increase in weight of fly larva from the time of hatching until pupation. Under optimal conditions the weight can increase to 90 milligrams.

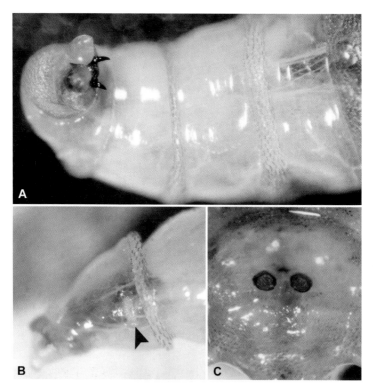

Fig. 8 External anatomy of a blowfly maggot. A: Head segment with protruding mouth-hooks and sensory organs; each larval segment is divided by a ring of annular spicules B: Anterior breathing apperture (anterior spiracle). C: Posterior spiracles.

ly, to ingest a large quantity of food as quickly as possible, storing some of it for later use during the migratory and pupal stages. In its short life span of nonstop feeding, a single maggot can process as much as 0.3 g of nutrient substrate, or necrotic tissue, or pus and wound fluid.

The maggot's intestinal tract is ideal for the optimal resorption and utilization of nutrients. In the case of *P. sericata*, the larval intestine

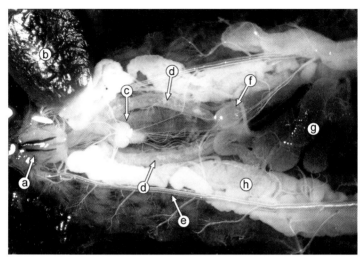

Fig. 9 Internal anatomy of a blowfly maggot. (a) Cephalopharyngeal skeleton with mouth hooks (sclerite); (b) crop; (c) ganglion (central nervous system); (d) salivary glands; (e) tracheae (respiratory system); (f) cardia (proventriculus); (g) midgut (intestine); (h) adipose body.

is five times the body length of the maggot. Ingested nutrients pass through the intestine at a rate of several millimeters per minute. This high metabolic capacity is reflected by the maggot's enormous growth rate. Under optimal conditions, a blowfly maggot can increase its weight one hundred-fold within a few days.

The maggot's energy stores are essential for fueling the process of metamorphosis into an adult fly, which occurs during the inactive pupal phase. The maggot also has another remarkable capability that is useful during metamorphosis: auto-disinfection, that is, the ability to rid itself of bacteria and other harmful micro-organisms. Bacteria ingested during the growth phase must be eliminated before pupation because they could otherwise multiply, infect, and kill the pupa.

3 Maggots and Wound Healing

History of Maggot Therapy

Fig. 10 Woodcut from *Hortus Sanitatis*, 1491.

My body is covered with worms and scabs,
My skin is broken and festering.

(Old Testament, Job 7:5, NIV Version, 1984)

The benefical effects of maggots on wounds have been recognized for centuries. Australian Aborigines used maggots to clean wounds for thousands of years. Military physicians stationed in Burma during World War II observed the medical use of maggots by the local residents. The Burmese traditionally placed maggots in wounds and covered them with mud and wet grass. Historic records indicate that Mayan Indians wrapped wounds with cloths soaked in animal blood and dried in the sun. The pulsating activity of maggots was soon observed beneath the wound dressings.

Military surgeons frequently observed and described the benefits of maggots infesting the wounds of fallen soldiers. The best known accounts are those of the French military surgeon Ambroise Paré (1510–1590), who reported on the Battle of Saint Quentin (1557), and the French surgeon Baron Dominique-Jean Larrey (1766–1842), who described his observations treating Napoleon's injured soldiers during an expedition to Egypt (1799).

Although his accounts tended to emphasize the negative, destructive nature of maggots, Ambroise Paré also described remarkably quick recoveries in his maggot-infested patients. However, Paré did not attribute this to the maggots, nor did he realize

Fig. 11 Ambroise Paré (1510–1590), described quick wound healing in maggot-infested soldiers.

that the "worms" were, in fact, fly larvae. Like most of his contemporaries, Paré believed that maggots developed spontaneously as part of the putrefaction process of devitalized necrotic tissue.

Napoleon's surgeon, Baron Dominique-Jean Larrey, recognized that "blue fly" maggots only removed dead tissue, after observing their beneficial effects on wounds. Larrey and his medical staff attempted to convince the soldiers that the maggots had a beneficial effect on the wounds, removing necrotic tissue and accelerating the natural healing process.

The American Civil War

John Forney Zacharias (1837–1901), a Confederate army surgeon during the American Civil War, was probably the first physician to intentionally expose festering wounds to maggots. According to Chernin (1986), Zacharias wrote,

> "During my service in the hospital at Danville, Virginia, I first used maggots to remove the decayed tissue in hospital gangrene and with eminent satisfaction. In a single day, they would clean a wound much better than any agents we had at our command. I used them afterwards at various places. I am sure I saved many lives by their use, escaped septicemia, and had rapid recoveries."

During this same time in the Northern hospitals, Union soldiers were having their wounds cleaned and dressed frequently, while the wounds of Confederate soldiers were left unkempt and maggot-borne. Reportedly, the flyblown wounds of the Confederate soldiers healed more quickly than those of the Union soldiers, and the Confederate soldiers were more likely to survive their wounds than were the Union soldiers.

William W. Keen (1837–1932), a Union army surgeon, noted the presence of maggots in wounds and commented that, in spite of their revolting appearance, the maggots evidently were not detrimental to the healing process (Goldstein, 1931). Still, the therapeutic application of maggots was hardly ever practiced by Union surgeons. The Confederate surgeons, on the other hand, often did not have any other choice since they rarely had enough dressing materials at their disposal (Adams, 1952). Joseph Jones (1833–1896), a Confederate doctor, wrote:

"I have frequently seen neglected wounds…filled with maggots…as far as my experience extends, these worms only destroy dead tissues and do not injure specifically the well parts." (*Cited after: Cunningham, 1970*)

The germ theory put forward by Robert Koch and Louis Pasteur in the 1880's revolutionized medicine in that it introduced the novel concept of hygiene-oriented practice, which forbade doctors to allow bacterially contaminated products to get anywhere near an open wound. Likewise, the development of antiseptic techniques completely reformed the field of surgery. Therefore, no physician practicing in the late 19th and early 20th century was willing to publicly propagate the idea of using unsterilized maggots.

World War I

William S. Baer, an American surgeon stationed in France during World War I, cared for two soldiers who lay injured on the battlefield for seven days before being discovered. By then, their wounds were teeming with maggots. Baer removed the maggots and observed, much to his surprise, that the wound beds were not filled with pus, but rather with healthy, newly formed granulation tissue. At that time, the mortality rate for soldiers similarly injured was 70 %. Baer recounts:

> "For seven days they lay on the battlefield without water, without food, and exposed to the weather and all the insects which were about that region. On their arrival at the hospital I found that they had no fever and that there was no evidence of septicemia or blood poisoning. Indeed, their condition was remarkably good, and if it had not been for their starvation and thirst we would have said they were in excellent condition...On removing the clothing from the wounded part, much was my surprise to see the wound filled with thousands and thousands of maggots, apparently those of the blowfly. These maggots simply swarmed and filled the entire wounded area. The sight was very disgusting and measures were taken hurriedly to wash out those abominable looking creatures. Then the wounds were irrigated with normal salt solution and the most remarkable picture was presented in the character of the wound which was exposed. Instead of having a wound filled with pus, as one would have expected, due to the degeneration of devitalized tissue and to the presence of the numerous types of bacteria, these wounds were filled with the most beautiful pink granulation tissue that one could imagine....These patients went on to healing, notwithstanding the fact that we removed their friends which had been doing such noble work."

After the war, as a professor of orthopedic surgery at Johns Hopkins University, Baer recalled his war experience while treating his many patients with chronic bone infections (osteomyelitis). He reared blowfly maggots, and applied them to the wounds of several patients. The foul smell of the wound disappeared along with the pus, and the symptoms of chronic inflammation subsided. All of these patients recovered and were discharged after just two months of treatment. To minimize the aversion of the patients and medical staff alike, and to keep the maggots from escaping, the physicians developed cage type dressings. To reduce the tickling sensation of the maggots moving around on healthy skin, Baer covered the sensitive edges of the wound.

Some of the patients developed gas gangrene and tetanus during the course of therapy. Although it could never be proven that the maggots were the cause of these infections, still Baer be-

History of Maggot Therapy

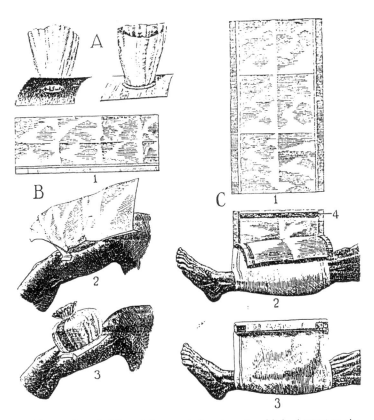

Fig. 12 A series of different "maggot-dressings," published 1934 in the *Journal of Bone and Joint Surgery*. Type A was used in cases of osteitis of the jaw and other small wounds, type B was used for almost any wound, and type C was employed for very extensive wounds of the limbs (Fine & Alexander, 1934).

lieved that sterile maggots should be used to prevent the spread of harmful micro-organisms.

Baer died in 1931, but not before convincing many of his skeptical colleagues of the scientific and therapeutic value of his

Council on Pharmacy and Chemistry

NEW AND NONOFFICIAL REMEDIES

The following additional articles have been accepted as conforming to the rules of the Council on Pharmacy and Chemistry of the American Medical Association for admission to New and Nonofficial Remedies. A copy of the rules on which the Council bases its action will be sent on application.

W. A. Puckner, Secretary.

SURGICAL MAGGOTS-LEDERLE.—Fly larvae of the species Phormia regina and Lucilia sericata. Marketed in bottles containing approximately 1,000 in a medium composed of desiccated hog's liver and 1 per cent nutrient agar.

Actions and Uses.—Surgical maggots-Lederle are proposed for use in treatment of chronic osteomyelitis and other suppurative infections; it is believed that the maggots clear away devitalized tissue after operation.

Dosage.—The wound is filled with maggots, which are allowed to remain for about five days, when they are flushed out with physiological solution of sodium chloride; the wound is then swabbed and a fresh supply of maggots is implanted. The average course of treatment is for six to ten weeks, the actual number depending largely on the size of the infected area and the individual response of the patient. Apparently, a substance is generated which becomes increasingly destructive to the maggots, since they can live in the wound only a few hours after several implantations. The antagonistic reaction varies in different patients. The product is forwarded to physicians according to a schedule which is designed to insure an active product.

Manufactured by the Lederle Laboratories, Inc., Pearl River, N. Y. No U. S. patent or trademark.

Fig. 13 Advert for "Surgical Maggots-Lederle" as it appeared in the *Journal of the American Medical Association* (JAMA), 1932.

work. His results with maggot therapy were so dramatic that his research and therapy were continued by many in the US and Europe. Maggots were regularly used for wound care in more than 300 hospitals in the United States alone, and over 100 publications on the subject appeared in the medical literature between

1930 and 1940. The pharmaceutical company, Lederle Laboratories, produced sterile maggots commercially for those hospitals that did not have their own maggot-breeding facilities (insectaries).

Penicillin was discovered by Alexander Fleming in 1928 but did not become commercially available until 1944. After that, the use of maggots in wound care rapidly declined and fell into oblivion, possibly due to the availability of antibiotic alternatives, to the decrease in prevalence of chronic infections, to the improved surgical techniques developed during World War II, or most likely to a combination of all three.

The "Revival" of Maggot Therapy Today

During the 1970's and 1980's, maggot therapy was used very infrequently, and only as a last resort for the most recalcitrant of infections (Horn 1976; Teich & Myers, 1986). In 1990, studies began to address the question: How does maggot therapy compare to other wound care methods currently in use? After all, why wait until everything else has been tried and has failed before using maggot therapy to treat a non-healing wound? The chance of cure is lessened by the very passage of time itself, so if maggot therapy is effective enough to use for the toughest of wounds, perhaps it should be used for less serious wounds, before they progress.

In a series of prospective and retrospective studies conducted between 1990 and 1995, it became apparent that maggot therapy was indeed more effective in debriding many non-healing wounds, and led to enhanced wound healing (Sherman et al., 1991, 1993, 1995; Sherman, 2002, 2003). Since then, the efficacy, simplicity, and low toxicity of maggot therapy has led to rapid acceptance by many wound therapists throughout the world. The International Biotherapy Society was founded in 1996 as a professional organization to advance the use, understanding, and acceptance of maggot therapy and other medical treatments that use live organisms. The first annual meeting of Biotherapists and Biotherapy research began that same year. In the United States, the non-profit Bio Therapeutics Education and Research (BeTER) Foundation now helps support patient care, education, and research in maggot therapy. By the year 2002, maggot debridement therapy (MDT) was again being used in over 2,000 health care centers world-wide. Having spent the past 50 years advancing their knowledge, skills, antibiotics, and surgical tools, medical practitioners have now invited therapeutic maggots back into the hospitals, clinics, nursing homes, and bedrooms of the wounded.

How the Maggots Work

After several decades of study, it has become increasingly clear that the observed efficacy of maggot therapy in acute and chronic wound infection is not attributable to any single substance but, rather, to a combination of different factors acting in synergy. Applied to infected wounds the maggots have three modes of action:

- They debride (clean) the wounds,
- they kill micro-organisms, and
- they stimulate wound healing.

"Maggot active principle...is not simply a combination of picric acid and calcium carbonate to form calcium picrate, as described by Stewart, nor allantoin, as described by Robinson. This has been proved by clinical trial, and can be understood better when we realize that maggots are fly embryos and, therefore, of necessity are rich in complex organic substances which, because of their embryonic nature, are growth stimulating."

(Livingston, J Bone Joint Surg. 18:751–756; 1936)

Debridement

Fly larvae secrete their digestive juices into their environment, liquifying the necrotic (dead) tissue, before it is then ingested by suction. This feeding technique does not require any type of biting apparatus. Wound tissue is not dissolved indiscriminately; the digestive enzymes of medical maggots appear only to dissolve necrotic tissue, not viable (live) tissue. The mouth hooks and spicules (fine, dorsally projecting, hook-like appendages (spicules) on each body segment) stimulate the wound tissue as the maggots crawl around, and facilitate the entry of the digestive enzymes into the tissue, and probably also trigger the release from host tissue of chemical mediators (cytokines) of wound healing.

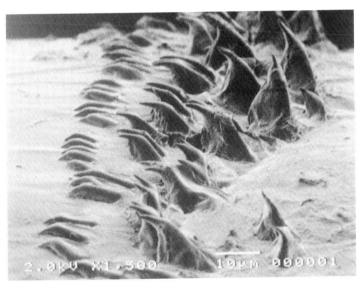

Fig. **14** Spicules of the fly larvae as seen under a scanning electron microscope.

3 Maggots and Wound Healing

Antimicrobial Action

Maggots disinfect wounds by mechanically removing bacteria and by releasing proteins that kill micro-organisms. Wound surfaces and necrotic tissue harbor large populations of microbial flora. Wound debridement by the maggots eliminates most of these micro-organisms. The remaining microbes and their toxins are flushed from the wound surface by the large quantities of fluid produced by the host and by the larvae during therapy.

Maggots have still further modes of antimicrobial action that developed as part of their evolutionary adaptation to life in rotting organic matter. This explains, in part, the incredible successes of maggot therapy in the treatment of wound infection. Given that the natural habitat of blowfly maggots (corpses, wounds, feces) is contaminated by an abundance of toxic, and lethal micro-organisms, it is not surprising that the maggots evolved a number of effective mechanisms for killing, eradicating, or managing these pathogens.

Blowfly larvae produce specific antimicrobial peptides that kill bacteria, probably by disrupting their cell membranes. Many of the digestive enzymes also probably kill microbes by destroying their protein surfaces. Furthermore, the maggots make the wound inhospitable for many microbes by manipulating the local environment (i. e., alkalinizing the wound bed) with their secretions of allantoin, ammonium bicarbonate, and urea.

Just as the composition of human digestive juices changes with the different food we eat, so too do the digestive juices secreted by larvae adapt to the special features of their external milieu. Research findings suggest that the concentration of bacteria in the surroundings affects the antibiotic effect of the fluids the maggots secrete. Other scientific studies have shown that maggots utilize the enzymes produced by certain bacteria to their own advantage. For example, they use their bacterial flora to better exploit their food source as is known to occur with the intestinal microflora of humans.

The symbiosis of maggots with certain bacteria is also of major relevance for the treatment of wound infections. The bacterium *Proteus mirabilis*, for example, secretes antibacterial toxins (including phenylacetic acid and phenylacetaldehyde) that kill other microbes but do not harm the maggots. Therefore, paradoxically, the bacteria can help eliminate other micro-organisms.

Antibiotics do not have a detrimental effect on the development of fly larvae. Therefore, maggots and antibiotics can be used concurrently in certain cases.

Stimulation of Wound Healing

Surprisingly, wounds treated with maggots have been found to heal faster and rapidly fill with healthy granulation tissue. Although some early researchers believed that the larvae merely facilitated the normal healing process by eliminating necrosis and infection, it has now been shown that the larval digestive juices actually contain growth factors (e. g., allantoin, ammonium bicarbonate, urea). These growth factors stimulate the healing tissue to grow faster and enhance the local delivery of oxygen (tissue oxygenation).

The validity of the assumption that mechanical stimulation by maggots traveling back and forth across a wound surface is the sole stimulus for wound healing has been severely challenged by the findings of studies using modern maggot dressings that physically separate the maggots from the wound. Separating the maggots from the wound with a membrane, thus allowing the secretions to reach the wound bed but preventing any physical contact between the maggots and the wound, still results in remarkably enhanced wound healing. After the application of maggots the three phases of wound healing, which normally take almost a year to evolve, begin early and proceed almost simultaneously. The maggots promote wound healing and remodeling by stimulating the growth of granulation tissue that rapidly fills the defect.

This simultaneously stimulates the growth of a durable epithelial cover over the wound and promotes wound contraction, leading to a rapid decrease in the size of the wound.

Primary mechanisms of Maggot Therapy

I. Debridement
- The maggots remove necrotic tissue by secreting digestive enzymes and by mechanically "scraping" the wound with their mouth hooks and fine spicules.

II. Antimicrobial action
- The maggots "flush" the wound by greatly increasing fluid production in the wound.
- The maggots secrete ammonia, calcium bicarbonate, and allantoin.
- The pH of the wound rises during maggot therapy.
- Large quantities of bacteria are killed in the acidic portions of the larval digestive tract (pH ~3).
- *Proteus mirabilis*, a bacterium in the digestive tract of maggots, produces substances that develop bactericidal effects in an acidic environment (since sterile maggots are used, they must absorb the bacterium from their surroundings).
- Laboratory studies have shown that the fluid secreted by maggots potently kills a number of micro-organisms (including MRSA and pathogenic *Streptococcus* strains).
- The maggots activate white blood cells while stimulating wound healing.
- The larval immune system contains antimicrobial substances (defensins).

III. Stimulation of wound healing
- The maggots mechanically stimulate wound tissue.
- Allantoin, urea, and ammonium bicarbonate secreted by the maggots promote wound healing.
- The fluid secreted by the maggots promotes cell growth and cell division.

- Insect hormones have also been found to stimulate the growth of cell cultures in laboratory experiments.
- The maggots secrete messenger substances (cytokines like IFNγ and IL-10) that act on human cells.
- The maggots lead to generalized activation of the host's immune system.

Medicinal Uses

It would be foolish to believe that maggots can heal any wound. Nonetheless, maggot therapy heals many wounds more rapidly than any other method, achieving good results in cases where other forms of wound treatment have failed. Because of the extremely low risks involved, it is sometimes justified to try maggot therapy even in cases where the response to treatment is uncertain, for example in chronic arterial occlusive disease. Of course, delaying definitive treatment is the most likely potential risk, so maggot therapy should not be continued for more than two weeks without seeing significant improvement. In patients with chronic wounds, the cause of the underlying disease must always be treated. The more comprehensive the interdisciplinary approach to treatment, the higher the quality of wound management.

Patients with acute and chronic wound infections, especially those with diabetic wound healing disorders, are good candidates for maggot therapy. The efficacy of maggot therapy in the management of treatment-resistant wounds such as diabetic foot ulcers, infected pressure sores (decubitus ulcers), and burns has been amply documented. However, the results achieved when using the maggots for treatment of stage 4 arterial occlusive disease (AOD) often are unsatisfactory.

In recent years, maggot therapy has been successfully used for treatment of Fournier's gangrene, which is still associated with a high mortality rate. In veterinary medicine, some encouraging preliminary results of maggot therapy have also been published.

Production of Sterile Maggots

Decades of successful therapy with *Phaenicia sericata* demonstrate that this species is well suited for the purpose, being relatively safe and effective. Experience with other species is very limited, but should be investigated so that local species can be used in each community whenever possible (Sherman et al., 2000). Arguably, laboratory strains that have been raised for therapeutic purposes may have an advantage over wild strains, given that strain differences are known to exist. Established colonies have a demonstrated track record of safety and efficacy, and also laboratory strains may have already undergone several generations of selection pressure eliminating or suppressing the more objectionable traits from the colony. Fly larvae ingest contaminated food from their immediate surroundings soon after hatching and can no longer be sterilized. However, the embryo inside the blowfly egg is sterile, and the membrane (chorion) enveloping the egg is extremely resistant. Therefore, fly eggs can be sterilized by disinfecting the surface of the egg and allowing the sterile larvae to hatch in a sterile container with sterile culture media. The optimal disinfectant should have high antibacterial potency but a low level of egg toxicity.

The variety of sterilization techniques described in the medical literature leaves the breeder free to select a preferred method of sterile maggot production. A microbiological laboratory must routinely test the sterilized eggs and hatched larvae for sterility within 24 to 48 hours. Until the microbiological test results are available, the maggots can be refrigerated at low temperatures to slow the further development of the larvae and to increase their shelf life. After hatching, the disinfected maggots will survive only 3–7 days, depending primarily on the adequacy of air

Fig. 15 Sterile maggots and empty egg capsules on the culture medium.

moisture, temperature, and food. Transportation often shortens their survival because of temperature and pressure extremes or insufficient oxygen or nutrients. Refrigeration can improve their survival considerably.

Myiasis

Maggot-induced wound debridement is really a controlled, therapeutic myiasis—a maggot infestation, administered according to rigorous standards, and closely monitored for safety and efficacy. Most of the potential complications of maggot therapy can be anticipated, and prevented, by understanding the nature of wound myiasis.

Occasionally, wild flies deposit eggs or live larvae on the skin or in the body orifices of a live human host. This can be called natural or non-therapeutic myiasis. Naturally occurring myiasis of the skin and soft tissue can be invasive (e. g., causing furuncular myiasis) or non-invasive (i. e., not entering or harming healthy tissue). Myiasis-causing flies can be classified as obligate or opportunistic (facultative). Sometimes blow flies, flesh flies, and other opportunistic myiasis-causing flies deposit their progeny on the wounds of live hosts instead of on their more customary substrate: cadavers or other decomposing organic matter. In general, the larvae of opportunistic myiasis-causing flies are non-invasive since they are used to feeding only on dead and dying organic matter.

It is essential that naturally occurring myiasis not be confused with, nor treated similarly to, therapeutic myiasis. Although natural wound myiasis has often been associated with beneficial outcomes (which is, after all, the observation that led to the original trials of maggot therapy for wound healing), nevertheless, wound healing is not always seen with natural myiasis. It is often not possible to determine the species of wild maggots involved in myiasis during the infestation, and thus to assess their invasive potential. In addition, non-disinfected maggots may harbor pathogens even worse than those already infecting the wound. Therefore, when patients are found with maggot-infested wounds, the maggots should be removed immediately. If it is determined afterwards that the patient might benefit from maggot therapy, then a course of treatment can be applied using disinfected larvae of proven efficacy and safety (that is, medical grade maggots).

When approaching a maggot-infested wound, it is important to keep the following items in mind:

1. Ensure the safety of the patient. Respiratory and cardiovascular integrity must be maintained at all times.
2. Remove the maggots, preserve them, and save them for subsequent identification, if necessary.
3. Save a few live maggots so they can be grown to adulthood; adult flies are easier to identify than larvae.

4. Interview the patient and closely inspect the patient and the facility so that contributing factors can be identified and mitigated.

When natural myiasis occurs in hospitalized patients, it is also incumbent on hospital staff to identify the precipitating factors in order to preclude such nosocomial infestations from reoccurring. In addition, a team of medical, nursing, facilities, pest control, environmental health, food service, and risk management/administration personnel should be assembled quickly to identify any contributing factors, and to recommend appropriate protocol changes. Nosocomial myiasis is generally dependent upon one or both of the following factors: 1) increased fly populations; 2) increased susceptibility of the patient. Factors that increase fly populations include: open, unscreened doors and windows; refuse left uncovered, or left for extended periods of time, or left near doorways, thereby attracting flies; food left on the ward or at patients' bedsides; open storage bins of soiled laundry. Factors that increase host susceptibility include: immobile or unconscious patients; incontinence; intubation; soiled dressings; malodorous secretions or fluids from any wound or orifice; wounds left uncovered, especially in patients who go outside to smoke or socialize.

In summary, although naturally occurring wound myiasis is often of some benefit to the infested patient, it is not an acceptable practice to permit its occurrence nor facilitate its continuation once it is discovered. Any patient thought to require maggot-induced wound debridement should be provided with medical grade maggots under the appropriate, controlled conditions needed to ensure optimal efficacy and safety.

4 Clinical Application of Maggots

The Classical Free-Range Maggot Dressing System

The edges of the wound should be protected from the maggots' digestive enzymes, from the infectious, necrotic wound drainage, and from the tickling or itching sensation caused by the maggots crawling over normally innervated skin. This is best done by covering the wound edges with strips of self-adhesive hydrocolloid, or cutting a hole the size of the wound in a sheet of hydrocolloid before placing it over the wound (Sherman, 1997). Zinc paste can then be applied to remaining areas of exposed skin to avoid irritation.

The next step is to place the young, 2–4-mm-long larvae on the wound. The recommended dose is 5 to 8 larvae per cm^2 of wound

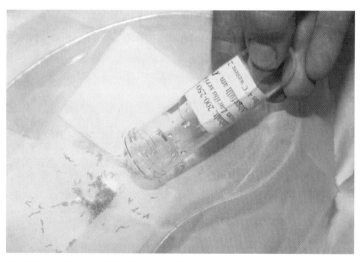

Fig. **16** A container of first to second instar larvae ready for use in therapy.

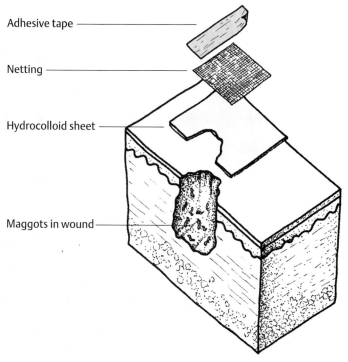

Fig. 17 The classical free-range maggot dressing.

surface area. The maggots can be rinsed out of the container onto a piece of nylon net, using saline solution, and then transferred from the net to the wound site either by using closed forceps or a swab, or simply by inverting the net over the wound site. Some manufacturers supply the maggots already on sterile gauze which can simply be removed from the vial, cut to size, and placed on the wound. Nylon netting, Dacron chiffon or other similar fine porous material is then glued and/or taped on top, completing the "primary" cage-like dressing. Refrigerating the maggots beforehand will slow them down, else they may escape from the wound as fast as you put them there.

The last step is to cover the primary dressing with a simple light gauze bandage ("secondary dressing"), which functions to absorb exudate and liquefied necrotic tissue during the treatment period. The secondary dressing can be readily changed as needed without letting the maggots escape. Moreover, the activity of the maggots can be observed through the translucent cage dressing whenever the secondary gauze dressing is removed.

The maggots are left in place for 2 to 4 days, after which they can be removed from the wound by peeling back the primary dressing while wiping up the maggots with a wet gauze pad, sandwiching the maggots between the dressings and the gauze pad. Alternatively, the maggots may be irrigated from the wound. The dressings should be discarded as soiled infectious waste ("redbag waste"), making sure to seal the bag containing the maggots in order to prevent their escape while they await autoclaving or incineration

The Containment-Bag Maggot Dressing System (Biobag)

Allowing maggots to crawl freely on the wound surface is not optimal for some wounds. Sometimes it is preferable to prevent the maggots from roaming on fragile tissue, or into a deep cavity (i. e., the pleural space or the nasopharynx) just beneath the necrotic wound. Furthermore, direct exposure to the maggots' sharp mouth hooks and spicules can be painful for some patients.

Also, it takes more time, effort, and training to properly dress a wound using free-range maggots as opposed to maggots confined in bags. In light of these considerations, some maggot therapists prefer maggot-containment dressings for delivering maggot debridement therapy. Contained maggots have been shown to be an

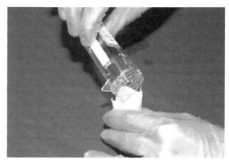
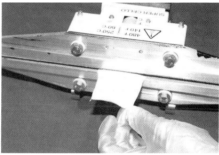

Fig. **18** Preparation of a Biobag.

effective method of maggot debridement therapy (Grassberger & Fleischmann, 2002). Two major designs are being used, both of which completely enclose the maggots. The Biobag (Polymedics, Belgium) is composed of polyvinylalcohol, an open-cell polymer;

disinfected maggots are heat-sealed within this four-sided pouch. Maggot secretions flow freely out of the Biobag and into the wound; the Biobag absorbs the liquified necrotic wound drainage, nourishing the maggots within. Polyvinylalcohol is also associated with wound cleansing and healing (Mutschler et al., 1980). Some containment systems are composed of nylon or Dacron chiffon. These sheer fabrics allow free passage of the maggot secretions and wound drainage, and also allow the maggots within to come into limited direct physical contact with the wound. Unlike polyvinylalcohol, however, these fabrics do not actively absorb wound drainage nor are they themselves associated with wound healing. Both containment designs offer the advantages of easy application and removal, decreased discomfort, decreased opportunity for maggots to escape, and the esthetic satisfaction of not having the maggots "loose" on the wound.

It has been argued that maggot containment systems, by limiting or preventing direct contact between the maggot and the wound, are less effective and efficient than maggot debridement therapy with free-range maggots (Thomas et al, 2002) because the host does not have the mechanical benefit of the maggots active mouth hooks and spicules. A clinical trial comparing the different containment systems with free-range maggot therapy clearly is needed to better evaluate and compare these different approaches.

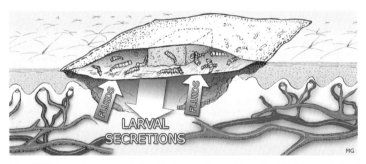

Fig. 19 Schematic view of the Biobag system. Absorbed fluids consist of liquefied necrosis, pus, bacteria and wound exsudate.

> Irrespective of the technical qualities of the different dressing systems used, the success of maggot debridement therapy is dependent upon an appreciation of the fact that maggots are living creatures. They can suffocate, drown, dessicate, or starve to death. They must be treated with care, and sensitivity.

Indications for Maggot Debridement Therapy

The treatment of acute wounds is straightforward, and fundamentally has been the same for thousands of years: stop the bleeding, and then protect the wound from harmful micro-organisms and environmental hazards. Usually, the body will heal itself quite nicely under these conditions.

When the body's ability to heal the wound is impaired, or under adverse environmental conditions, then the wound may become chronic. Chronic wounds put the patient at continued risk of infection, which may impair the body's capacity to heal even further. Chronic wounds result from, and may themselves cause, further imbalances in the biochemical mediators (cytokines) that regulate and promote wound healing.

The Diabetic Wound

Three common errors, individually or collectively, frequently lead down the fateful path to amputation: diabetes management, lack of diligence on the part of the patient in monitoring and adjusting the blood glucose, and negligence in the imperative, meticulous examination of the feet. Both physician and patient must examine the feet at every opportunity. Most amputations in diabetic patients could be avoided by local preventive measures, such as wearing comfortable orthopedic shoes, the prevention or immediate medical treatment of all injuries, and surgical correction of skeletal foot anomalies to improve the distribution of pressure across the foot.

4 Clinical Application of Maggots

Chronic non-healing foot wounds in diabetic patients are characterized by an absence of natural wound healing processes. Progressive tissue death tends to occur, resulting in the loss of toes (diabetic gangrene) or the destruction of skin on the soles of the feet and in the underlying layers of padding tissue (malum perforans). Dead tissue immediately breeds bacteria and fungi. The breakdown of proteins leads to the characteristic stench of necrotic tissue. Microbial invasion of the neighboring healthy tissue may lead to further necrosis, cellulitis, lymphangitis, bacteremia, osteomyelitis, or death.

Many diabetic patients with foot wounds are ideal candidates for maggot debridement therapy (Mumcuoglu et al., 1998; Fleischmann et al., 1999; Sherman, 2002). The maggots gently and thoroughly remove necrotic tissue by mechanical action and by enzymatic liquifaction (proteolytic digestion). The maggots secrete antimicrobial peptides into the wound, kill ingested bacteria in their gut, and alkalinize the wound. Furthermore, the increased production of wound secretions rinses bacteria and bacterial toxins out of the wound. Maggot secretions promote wound healing and often dramatically shorten the wound healing time, provided that blood flow in the major arteries is adequate to support healing.

Case No. 1

This 66-year-old man suffered from a diabetes-related foot deformity (Charcot's syndrome). As a result of bone fragments exerting pressure on blood vessels in the nearby soft tissues, the great toe of the right foot became ischemic and the toe had to be amputated. The surgical wound became infected, leading to further surgeries, progressive enlargement of the wound and, ultimately, to loss of the first and second metatarsal bones. Despite six months of inpatient clinical and surgical management, the wound did not show any signs of healing. In fact, the condition of the wound progressively worsened. The physicians and the patient jointly determined that amputation of the right lower leg was needed to relieve the patient's suffering.

However, maggot debridement therapy was attempted as a last resort to salvage the foot.

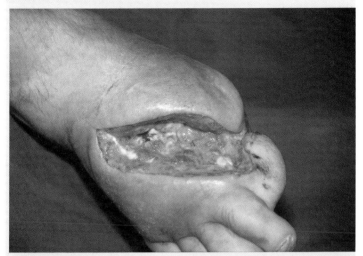

Fig. 20 Diabetic gangrene before maggot debridement therapy.

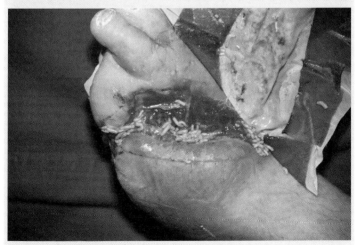

Fig. 21 Same patient: appearance of wound before removal of the maggots.

The maggots were placed directly on the wound surface and covered with nylon net. The netting was applied loosely enough to provide the maggots with plenty of growing space and air, then taped securely to contain them. The dressing was removed three days later. By then, the maggots had grown to a length of 8 mm and the previously unresponsive wound bed had become red with healthy blood flow and granulation tissue. After more treatment cycles, the patient was able to place weight on the foot without pain. He was then discharged from the hospital for further outpatient care. The wound completely healed within six weeks, and the patient was able to return to work at his gardening business.

Comments:

If maggot debridement therapy did such a good job healing this wound, thus preventing amputation, imagine how much sooner it would have healed, and how much less destruction there

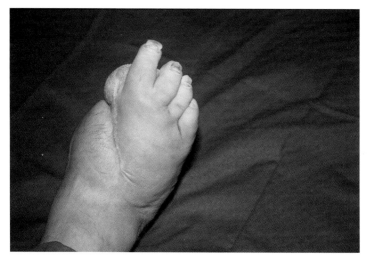

Fig. 22 Same patient. final result after wound healing.

would have been, if this patient had been treated with maggot debridement therapy earlier in the course of his non-healing wound!

Chronic Leg Ulcers

Chronic ulcers of the lower leg can develop for a number of reasons. The most common cause is impaired blood flow, for example an insufficient arterial blood supply to the limb or failure of venous return to the heart. Mixed ulcers have both arterial and venous components.

The etiologic diagnosis of chronic leg ulcers is difficult and often costly. However, specific therapy requires an exact diagnosis. If the arterial supply to the limb is deficient, the goal of treatment must be to improve the flow of the blood and to restore arterial patency. If compromised venous return has led to vascular congestion, leg

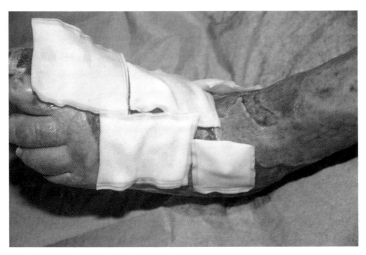

Fig. **23** Biobags applied to foot wounds.

compression (compression bandages or elastic stockings) must be carried out, augmented by surgical measures as needed.

Maggot therapy may help heal the leg ulcer, but it will not resolve the underlying cause. Maggot debridement therapy should be performed only after the underlying circulatory disorder has been properly diagnosed and treated. Malignant neoplasms can arise from non-healing wounds that persist for several years. Therefore, it is imperative to examine the histology of tissue samples taken from the edge of the wound. Current findings indicate that medicinal maggots do not have a lethal effect on neoplastic tissue. Immediate surgical or dermatological treatment of tumors is therefore required.

A good knowledge of pathophysiology is required for successful treatment of non-healing leg ulcers that have persisted for many years. The process of wound healing is normally very effective and usually results in complete closure without any specific intervention. The acute response is hemostasis; the next step is the prevention of infection. By producing large quantities of wound fluids, bacteria and bacterial toxins are rinsed out of the wound, preventing them from colonizing the wound surface. Cellular and humoral immunity further protect the body from microbes. Within a few days, an extremely resilient layer of pink, vascularized granulation tissue develops, sealing and protecting the wound from harmful micro-organisms. The production of wound fluid now decreases, and new skin can be produced around the wound edges to achieve wound healing. Collagen is laid down, the wound contracts, and the edges are brought closer together as the skin regenerates. These phases do not occur in stepwise sequence, but rather as a continuum.

Serious consequences can arise from the prolonged failure of wound healing. Friable granulation tissue or poorly vascularized scar tissue may develop over the wound bed, blocking the absorption of essential nutrients from layers of healthy tissue below the wound, or preventing new skin growth or skin grafting.

Maggots cannot easily dissolve tendons, ligaments, bone, or large plates of scar tissue. Maggots fail at this task whereas a surgeon can eliminate the problem very quickly, easily, and inexpensively using a scalpel.

In spite of this limitation, the adjuvant use of maggots for treatment of leg ulcers can produce outstanding results, provided the maggots are assigned their specific niche in the overall treatment strategy and provided they are selectively employed to debride wounds and combat infections.

Maggots have been reported to promote some *Pseudomonas* and *Proteus* wound infections. Organisms in the wound should therefore be regularly isolated and identified. If evidence of a *Pseudomonas* or *Proteus* infection is found, the patient may need to be switched, at least temporarily, to another antimicrobial treatment method.

Case No. 2

This 60-year-old woman developed an extremely painful left lower leg wound one year ago. The wound expanded progressively despite regular dressing changes. After the first six months of treatment for her venous stasis disease, her femoral arteries were found to be obstructed. Arterial flow to the leg was surgically restored. Then maggot debridement therapy was initiated. Medical grade maggots were applied to clean and disinfect the leg ulcer, which was then covered with grafted skin. Biobags were applied to two other skin defects on the leg. When they were removed 4 days later, the wounds were clean and starting to heal. During treatment, the patient also received counseling to help her cope with the psychosomatic effects of the painful chronic wounds.

4 Clinical Application of Maggots

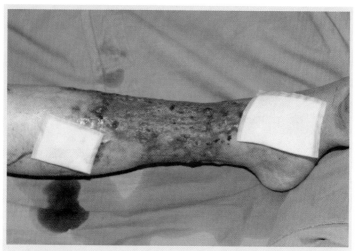

Fig. 24 Modern Biobags used to treat a leg ulcer of mixed etiologies.

Comments:

The cause of chronic wound infections can be multifactorial. The underlying cause must be treated in order to optimize the likelihood of wound healing. If seriously compromised arterial blood flow to a limb is not detected and treated quickly enough, the patient may lose the limb. Deficient blood flow not only encourages life-threatening wound infections, but also causes pain. If the pain becomes chronic, phantom limb pain may plague the patient even after the affected limb has been amputated.

Maggot debridement can cause significant pain, even when a Biobag is used. Therefore, some patients will require analgesia. Simple oral or parenteral analgesics are usually sufficient. If the pain still is so severe that it jeopardizes the completion of therapy, the administration of a peridural opiate block or plexus anesthesia via a catheter or other forms of pain therapy should be considered. Alternatively, the cycle of maggot therapy can be abbreviated, at which point the maggot-induced pain abates im-

Indications of Maggot Debridement Therapy

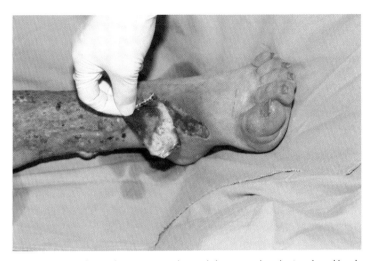

Fig. 25 Mixed ulcer. The maggots cleaned the wound and stimulated healing.

mediately. The analgesic should reliably eliminate the pain for several days, during which time surgical treatment of the wound, veins, or scars can be carried out if necessary. Vacuum therapy has been shown to accelerate wound healing after skin grafting.

Pressure Ulcers (Decubitus Ulcers)

The average life expectancy is gradually rising due to prosperity, good hygiene, and high-quality medical care in affluent countries of the world. People with multiple or incapacitating/immobilizing medical problems are being kept alive, but require prolonged periods in bed. Extended bed rest can alter circulatory integrity, and immobility often leads to local ischemia and soft-tissue necrosis in those parts of the body subjected to the highest pressure, for example the back of the head, shoulders, hips, sacrum, buttocks, and heels. Dietary deficiencies and impaired immunity can make patients more susceptible to wound infections and impaired healing; this often results in further nutritional deficiencies that lead the patient into a spiraling cycle of deteriorating health.

Preventive measures are essential for maintaining the patient's quality of life. Regular hygiene, a healthy diet, forced mobilization, frequent repositioning of the patient, and the use of special mattress systems for optimal pressure distribution reduce patient discomfort and prevent avoidable morbidity and hospital treatment costs.

If pressure ulcers do develop, maggot debridement therapy can be useful to clean and disinfect the wounds (Sherman, 2002). Once the maggots have debrided the ulcer, the wound might be closed surgically or naturally. Surgical measures can provide rapid wound closure and healing, especially in atypical pressure ulcers in younger patients with impaired pain perception due to a nerve/spinal-cord disease or injury.

Pressure ulcers of the buttocks or peraneum can be difficult to treat with free-range maggots in multi-layered cage dressings. These dressings frequently fall off because of sweating and soiling. Therefore, a containment bag can be very helpful to keep the maggots over the ulcer. If the patient rolls over or sits on the dressing, the maggots can be fatally crushed or may suffocate as a result of the high pressure. This can be prevented by inserting a sturdy cube of textile spacer material into the containment bag, which also ensures that the maggots have enough room to grow and produce their larval secretions.

Indications of Maggot Debridement Therapy

Case No. 3

This patient, an 80-year-old woman living at home with her niece, had a history of cardiac insufficiency and renal failure. A stroke that occurred one year earlier left her paralyzed on one side and bedridden. The patient was no longer able to care for herself.

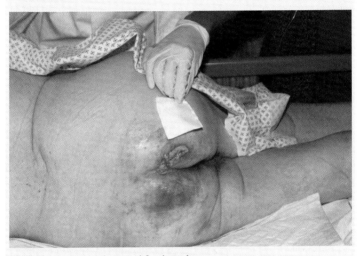

Fig. 26 Decubitus ulcer and fresh Biobag.

Despite daily home health care, a pressure ulcer developed on the patient's buttocks. The wound progressed and she developed life-threatening renal failure. By the time she was hospitalized, the pressure ulcer was covered by necrotic tissue and the surrounding skin was deteriorating. A Biobag was placed on the wound and loosely covered with moist compresses. After three weeks in hospital, her renal function stabilized and she was allowed to return home with a pressure-reducing mattress. Complete wound healing was achieved after four more cycles of maggot therapy.

Comments:

Small bedsores usually heal quickly after the application of maggots, provided that the patient continues to be repositioned frequently and that adequate measures are taken to improve the patient's general physical and nutritional health. A balanced diet and scrupulous personal hygiene are essential.

Case No. 4

This 70-year-old woman suffered from multi-infarct dementia. A metastatic breast carcinoma later developed, leading to a severe loss of muscle strength that left the patient bedridden. The home care nurses and the patient's family were unable to tend to the patient properly. The family doctor was consulted because of the fetid odor pervading the patient's room. The doctor identified a huge pressure ulcer in the buttocks region as the cause of the smell. The patient was admitted to hospital after an attempt at outpatient therapy failed. By then, the pressure ulcer had reached frightening proportions. It involved the entire gluteal region with deep muscle necrosis and exposed osseous structures of the sacrum. The patient's condition was so poor that the doctors did not expect the wound to heal. Nevertheless, it was decided to let maggots clean the wound to eliminate the offensive odor and to facilitate patient care, even though wound healing could not be expected.

Comments:

Decaying tissues release aeromatic substances with offensive odors. As a result, patients with incurable wound infections or decaying tumors tend to become isolated from society—avoided by friends, family, and health care providers—even though they are in great need of human contact and medical care. The warmth and security of the home must then be abandoned for the emotionally sterile environment of the hospital.

Even in some seemingly hopeless cases, maggot therapy can eliminate decaying tissue and its offensive odor.

Acute Wound Infection

In acute wound infection, the absence of a protective barrier of healthy skin allows bacteria to invade the wound. The immune system is specialized to kill micro-organisms previously encountered, and some not previously encountered.

The introduction of antibiotics 60 years ago has allowed humans to win the fight against micro-organisms that previously alluded our defenses. The indiscriminate use of huge quantities of highly potent antibiotics has led to increasing insensitivity of some bacteria.

Some evolutionary biologists already foresee the end of the brief "Golden Age" of antibiotic control of infections. It takes humans 1000 years to produce as many generations as bacteria can produce in a single day. Such short generation times allow bacteria to more easily and quickly adapt to changes in their environment. For example, penicillin killed all known *Staphylococcus* strains in 1941. The first penicillin-resistant strains had developed by 1944 (the bacteria secrete enzymes that destroy penicillin) and today, nearly 95 percent of all staphylococci are resistant to penicillin.

Fortunately, due to the constant development of new drugs and more sophisticated treatment procedures, we are still able to keep acute wound infections under control in most cases. However, rapid control of local infection requires the selective use of systemic antibiotics and resolute intervention with modern surgical weapons, e. g., vacuum-assisted wound closure (VAC) and maggot debridement therapy.

In view of the development of multiresistant strains of bacteria, maggot therapy may soon gain special importance in the prevention of nosocomial infections.

Because of the unavailability of timely medical treatment, disaster scenarios such as earthquakes, mass accidents, and armed conflicts are associated with high secondary mortality due to wound infection. The war surgeons of old observed that the spontaneous colonization of wounds by maggots promoted wound healing. Likewise, the prophylactic application of maggots in Biobags could help mass disaster victims survive through the critical period when proper medical care is unavailable.

Case No. 5

This 30-year-old woman was admitted to hospital with critical injuries, after crashing on her motor scooter. An external fixation system was immediately applied to stabilize an open comminuted fracture of her left lower leg. However, the soft-tissue injuries were so extensive that infection developed, causing high fever and hepatorenal dysfunction. Methicillin-resistant *Staphylococcus aureus* (MRSA) was identified as the causative organism. Multiple attempts were made to surgically remove all infected tissue. The patient's lower leg was ultimately shortened 14 cm, yet the infection remained uncontrolled.

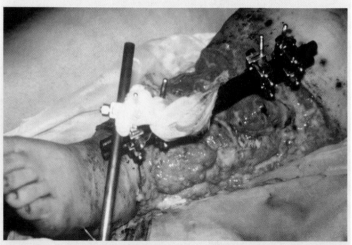

Fig. 27 Patient with acute wound infection before maggot therapy.

Above-knee amputation first seemed to be the only life-saving option, yet the team ultimately decided to try maggot debridement therapy in a last-ditch effort to salvage the limb. Therefore the patient was transferred to the intensive care unit of another hospital that was equipped to perform larval therapy.

The wound was found to be in wretched condition. The broken and shortened lower leg bone (tibia) jutted out of a pus-filled bed of necrotic muscle tissue. An unusually large number of sterile maggots (> 1 000) was applied to the wound. The maggots were contained in sterile netting, which was carefully fitted around the external fixation device and secured with tape.

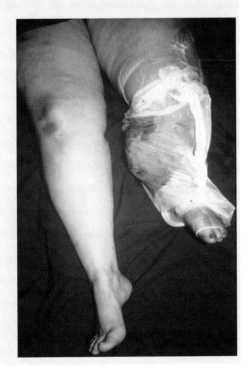

Fig. 28 Same patient: open-range maggot dressing.

The maggot dressings were replaced three times over the next two weeks. The patient's general state improved rapidly, and she was transferred to the sepsis ward. The following week, a wet sponge dressing (Biogard by Polymedics, Peer, Belgium) was applied for further wound cleaning. The wound surface was clean, disinfected, and well supplied with blood. Healthy skin from another part of the body was then grafted onto the wound, which subsequently healed without further complications. Finally, leg-lengthening surgery was performed to restore the leg to its original length.

Comments:

If the immune system is not working properly or is attacked by highly virulent pathogens, a wound infection can lead to death within a few hours to days. Gas gangrene and tetanus are the best known examples, but epidemic streptococcal infections can be just as deadly.

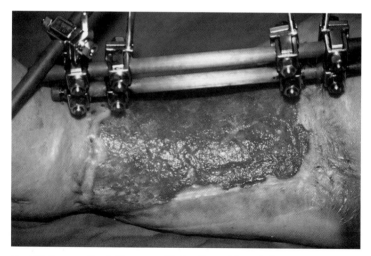

Fig. 29 Same patient: the wound is healing without inflammation.

Indications of Maggot Debridement Therapy

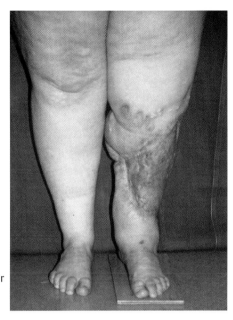

Fig. 30 Same patient after wound healing and leg lengthening.

The surgical approach to halting this life-threatening infection includes pathogen-specific, high-dose antibiotics and, most importantly, uncompromising excision of all infected tissues. In the extreme case, this could very well mean amputating a leg to save a patient's life. As Michael DeBakey, the American vascular surgeon, used to say, "Life before limb." Surgeons have followed this universal principle since time immemorial.

Despite the seriousness of this woman's injuries, it was fortunate that she survived the serious accident, that her infection (*Staphylococcus aureus*) was killed efficiently by maggots, and that she was hospitalized in a center where she could obtain the required therapy.

If the patient's wound had not responded as well and as quickly to maggot therapy, above-knee amputation would have been the only recourse. Of course, acute wound infections do not always respond to maggot debridement. Treatment must be individualized and must be closely monitored for efficacy. Septic wounds must be treated quickly and effectively, else the consequences can be grave. The indiscriminate use of maggots or any other wound treatment can have lethal consequences.

Chronic Wound Infection

When an infection becomes chronic, the patient and the causative organisms have reached a stalemate. The bacteria may be so deeply lodged in body tissue that they are largely safe from attack by antibiotics or the host's immune system.

Bacteria frequently invade the bone, where they form innumerable small foci of infection. Chronic bone infection can become a lifelong problem if the diseased bone segment is not completely resected. Episodes where fistulae and painful abscesses form may be followed by symptom-free remission periods. The chronically inflamed tissue may undergo malignant degeneration, and amputation of the leg above or below the knee is often the only hope for these mainly young patients. Careful attention to asepsis is imperative during primary care of accident victims and during surgical interventions, especially when the skeletal system is affected. This helps lower the number of acute infections and, thus, the prevalence of chronic infections.

The early pioneers of maggot therapy publicized the absolutely incredible results of maggot debridement therapy in patients with osteomyelitis during the early 20th Century. Meanwhile we are experiencing a renaissance of this biosurgical technique for the treatment of chronic bone inflammation. The use of maggot debridement therapy as an adjuvant to modern bone surgery, vacuum-assisted wound closure for high-dose local drug administration, and sophisticated plastic surgery techniques has raised the quality of wound management to new heights.

Utilization of the entire range of biosurgical possibilities, including modern dressing systems, suitable spacer materials, and maggot containment bags (Biobags), contributes to the success of maggot debridement therapy. Therapeutic success is enhanced by special microbiological knowledge and the use of state-of-the-art technology within the infrastructure of a specialized biosurgical unit.

Case No. 6

This 30-year-old woman had a hysterectomy on 18th January 2000. Infection of the wound with multiresistant staphylococci necessitated multiple surgical revisions. The surgeons could not immediately bring the infection under control. In the meantime, the patient developed a large wound through which the bladder and a loop of bowel could be seen. On 8th February 2001, the patient was therefore transferred to the surgical ward of a university hospital.

Although appropriate treatment measures were implemented, including an attempt to perform a sliding flap graft to close the wound defect, the infection remained active. The patient was discharged after it was decided that she could be treated on an outpatient basis. In late August, she awoke one night in a pool of blood. Her left pelvic artery had ruptured because of infection-related erosion. Immediate emergency surgery saved her life but, under these grave circumstances, it was necessary to tie off the affected pelvic artery to stop the bleeding, even though this seriously jeopardized the blood flow to the leg. Fortunately, the patient survived and her leg did not have to be amputated, although the blood flow was severely compromised.

4 Clinical Application of Maggots

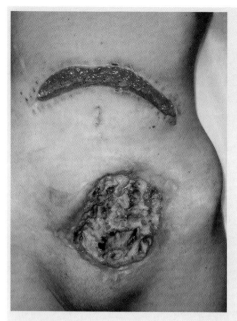

Fig. 31 Chronically infected abdominal wound: initial appearance.

The infection continued to spread. Above the exposed roof of the bladder, a small intestinal fistula was found to be leaking bowel contents into the wound. The attending physicians managed to close the fistula, but the prognosis was gloomy. They predicted that the patient would not survive more than a few weeks. In a last desperate attempt, the patient was transferred on 30th September 2001 to a distant hospital specializing in wound management procedures, including maggot debridement therapy.

Examination revealed a malodorous wound, coated with thick layers of decaying tissue and full of craters. Bladder-wall irritation was causing severe pain on urination. The abdominal wound, situated left of midline, ran past the bladder and projected down to the region of the pelvic vessels. The infected epigastric surgical wound was also visible.

Indications of Maggot Debridement Therapy

On 2nd October 2001 a Biobag filled with maggots was strategically placed over critical areas of the wound. To avoid unnecessary complications, no maggots were applied in the region of the former small intestinal fistula or vessel-containing craters. The Biobag was removed three days later. The bag was slit open for inspection purposes, revealing healthy, thriving larvae. The odor of decaying tissue was hardly perceptible. Inspection of the wound revealed significant debridement and first signs of healing. After two weeks of maggot therapy, all signs of infection had subsided. Granulation tissue proceeded to fill the wound bed, and new layers of healthy tissue began to cover the wound.

The patient's general condition improved tremendously. She regained her appetite. Urinary urgency and pain on urination subsided, and the size of the wound decreased continuously. The patient slowly regained confidence in her physicians and their treatment methods.

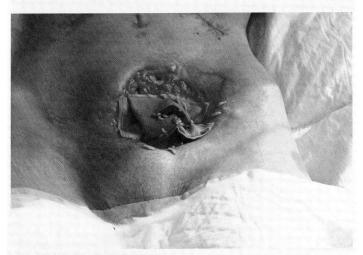

Fig. **32** Same wound: Biobag slit open to reveal the maggots inside.

4 Clinical Application of Maggots

> The maggots had done their job well; they had eliminated the infection, debrided the wound, and stimulated the healing process. The next step was to perform plastic surgery to achieve a functionally and cosmetically acceptable abdominal wall closure and to improve circulation to the leg by repairing the pelvic blood vessels.

Comments:
Maggots are especially effective in combating extensive soft-tissue infections caused by *Staphylococcus* or *Streptococcus*. W.S. Baer, the pioneer of maggot therapy, also observed a positive effect of maggots on gas gangrene and *clostridium*. He later performed animal experiments to confirm his observations. However, maggots may not adequately kill bacteria like *Proteus*, *Pseudomonas*, and *Escherichia coli*. Therefore, in addition to systemic antibiotics, an instillation therapy (VAC-Instill, KCI) may be needed for immediate cyclic application of antiseptics in wounds with severe sepsis.

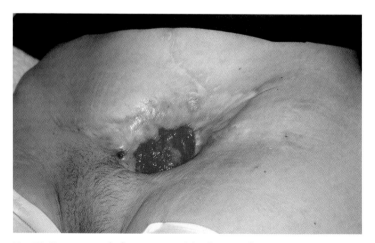

Fig. 33 Same wound after maggot debridement therapy.

The overall treatment strategy is based on the results of microbiological testing. Specific antibiotics and local treatment measures are selected based on the type of organisms found in the wound. This knowledge prevents the treatment from reaching a dead end.

The highly active digestive enzymes secreted by the maggots dissolve devitalized tissue, not healthy tissue. Maggot debridement can therefore be classified as a generally safe debridement method in humans. However, in the case of our patient, allowing the maggots to crawl freely in the abdominal wound would have posed an unacceptable risk of complications. Direct exposure of the visible and inflammation-damaged pelvic vessels or the former small intestinal fistula to the maggots (i. e., to their secretions and mechanical activity) could lead to the leakage of bowel contents or to hemorrhaging. The use of a maggot containment bag (Biobag) therefore increased the safety of maggot therapy.

Case No. 7

Using his cell phone while driving, this 38-year-old male caused a head-on collision after his car drifted into oncoming traffic on 19th August 1998. His severe head wounds and a comminuted left femur were treated at the nearest hospital. The patient's condition stabilized a few days later, but the surgical wound on his thigh became infected.

This marked the beginning of a year-long series of unsuccessful attempts at surgical treatment. Each surgery raised new hopes that were ultimately dashed. The infection eventually spread to the hip joint, and the head of the femur had to be removed on 22nd October 1998.

Then, yet another complication arose: the patient's left leg became dysfunctional. Moreover, the infection still was not under-control. A tragic climax was reached on 12th April 1999, when physicians at a university medical school told him that all attempts had been in vain. They felt that above-knee amputation could not be avoided in view of the lack of leg function and since the chronic bone infection had spread throughout the thigh by then. Also, all previous treatment attempts had failed. However, the patient had not given up hope; he clung to the slim hope that maggots might be able to resolve the infection.

On 19th April 1999, just prior to the trial of outpatient maggot therapy, the prognosis was grim. The infected femur could be seen in the depths of the 30-cm-long, malodorous, festering wound. The maggots would have to be carefully placed in the wound so that they would not drown in the profuse wound fluids or be lost between the deep, steep edges of the wound.

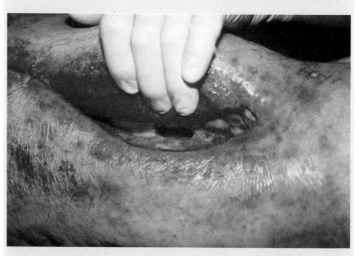

Fig. 34 Chronic inflammation of the bone (osteitis): initial appearance.

A spacer was devised to keep the wound open and well-aerated. The spacer and the maggots were securely enclosed in fine nylon mesh.

After several weeks of maggot therapy, the infection unexpectedly receded, the bone fracture healed well enough for the steel plates to be removed, and the wound fused together. Unfortunately, the patient fell down the stairs and broke his leg again on 5th June 2000; but this fracture healed, too, after it was surgically stabilized using an external fixation device. The excised hip joint was replaced with an uncemented titanium prosthesis on 20th June 2001. This was a milestone on the road to restoring the functionality of the patient's leg and, thus, his quality of life.

Fig. 35 Maggots inside the femoral wound.

Comments:

In spite of the astonishing treatment results observed in this case, it is unlikely that maggot debridement therapy resolved the underlying problem. Since the maggots cannot reach the microabscesses deep within the bone, rekindling of the infection can occur at any time.

4 Clinical Application of Maggots

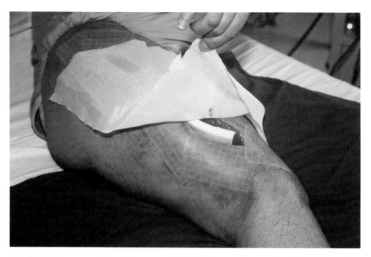

Fig. 36 Same wound with spacer to hold it open to aerate the maggots.

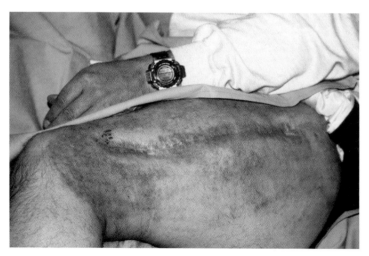

Fig. 37 Same wound after completion of healing.

In order to objectively assess the efficacy of maggot debridement therapy in osteomyelitis, an adequate number of patients must be followed and documented for a period of at least five years. This has not been done so far. Moreover, maggot debridement is rarely ever used as the only method of treating chronic osteitis; it is rather a single link in a chain of important surgical measures. It is not always possible to determine which link was responsible for the success of treatment. Nonetheless, in combination with surgery, maggot therapy has long been used to improve the treatment and outcome of this debilitating and sometimes fatal chronic disease.

Adverse Effects and Risks of Maggot Therapy

Pain and skin irritation are the most commonly reported side effects. Caution is advised when exposed loops of bowel or blood vessels contain necrotic elements, as the larval secretions may dissolve the devitalized tissue, resulting in the development of an intestinal fistula or hemorrhage.

Maggots produce an environment that is generally deleterious to bacteria. However, larval secretions do not inhibit the growth of some *Pseudomonas* or *Proteus* strains as they do other pathogens. Therefore, when these pathogens contaminate a wound, they may increase in number or even begin to infect the site as their microbial competitors are eliminated during maggot therapy. Although never reported in humans, high maggot loads in sheep are associated with the production of ammonia, high serum ammonia levels, altered consciousness, and even death. Anyone experiencing fevers, altered mental status, or other systemic symptoms while receiving maggot debridement therapy should have serum ammonia levels checked and blood cultures performed, along with the routine diagnostic and treatment procedures.

The hazard most feared by hospital administrators is often that escaping maggots will spread infection or mature to flies spreading throughout the facility. While theoretically possible, these scenarios are not the result of therapy but rather the result of multiple breaks in standard procedures. While medical grade maggots may start off as disinfected, and while they actively produce and secrete antimicrobial substances, maggots that have been used in wound care should not be considered "sterile." Indeed, they should be handled as fomites, and discarded with other infectious waste once they are removed from the patient. The waste bag should be hermetically sealed, and placed in a second, sealed, bag to ensure that the maggots cannot escape while awaiting disposal. The use of maggots within a secure dressing will help keep the larvae from escaping or maturing, and thereby minimizes the risks of cross-contamination.

Psychological complications of maggot therapy such as fear and anxiety are generally the result of misinformation. Therefore, the preventive intervention is to provide accurate and thorough information. Only in this way is it possible to dispel such common and frightening myths as: "medical maggots burrow under the skin," "maggots can multiply," and "the maggots can mature to adult flies while in the wound."

Perspectives

The culturally ingrained concept of the fly as the antithesis of health—or worse yet, as the embodiment of death itself—has prevented a wide acceptance of this age-old natural healing method. Yet, the use of disinfected *Phaenicia sericata* larvae in wound management has a solid history and an expanding literature to support its use. Treatment efficacy has been demonstrated repeatedly, and serious side effects are rare. Slowly but increasingly, medical grade maggots are being escorted back into hospitals and clinics throughout the Western World. If a patient and healthcare provider have no objections to using medicinal maggots, then we should help bring down all other barriers that stand in their way.

The introduction of maggot containment bags (e. g., the Biobag) has significantly increased the acceptance and simplicity of maggot therapy in Europe. Whether or not isolated larval secretions or pharmaceutically manufactured larval active substances will be able to replace conventional maggot therapy in the future remains to be seen. But for now, medical maggots are again a welcomed member of our modern, multidisciplinary, wound care teams.

Cost Effectiveness

Preliminary studies have shown maggot therapy to be a cost-effective method of debridement. In a prospective controlled study of maggot therapy vs hydrogel for venous stasis ulcers (Wayman et al., 2001), all six maggot-treated wounds were completely debrided after just one cycle of maggot debridement therapy, at a median cost of 78.64 GBP (approx. USD 133.00). Four of the six venous stasis ulcers not treated with maggot therapy were still necrotic four weeks later, at a cost of over 136 GBP (approx. USD 230.00) for each of the six wounds. Further cost-effectiveness studies are needed; but at the present time the low cost of maggots plus their acknowledged efficacy make them an economical treatment option.

5 Appendix

Frequently Asked Questions

What are the risks associated with maggot debridement therapy?
We did not find any evidence of significant risks or effects in conjunction with the clinical use of sterile larvae of the blowfly species *Phaenicia sericata*. However, one should remember that maggots are very efficacious therapeutic agents that should only be used by properly trained care providers according to the standard rules of good hygienic practice.

Although the potent enzymes produced by maggots only dissolve necrotic tissue, they are still capable of irritating healthy tissues, especially skin. This most commonly occurs when too many maggots are applied to a wound, or when they are left in the wound too long after successful debridement. Maggots should not be applied near damaged or exposed blood vessels.

Can maggots transmit infections?
When using disinfected maggots produced according to the rules of good clinical practice, there is no need to fear that the maggots will introduce harmful organisms into the wound. However, "wild" maggots can function as vectors of infection.

Do maggots burrow into living tissue?
Unlike some fly species, medical grade maggots neither attack nor burrow into living tissue. Furthermore, they cannot survive deep under the skin because they need air to breathe.

Can maggots mature into flies inside a wound?
A freshly hatched maggot requires at least seven to 14 days to complete the life cycle, at the end of which a new fly emerges. The larvae used in maggot dressings are renewed every two to

four days to ensure that the larvae cannot mature into pupae. Furthermore, maggots need a dry environment for pupation. When they mature, the maggots therefore attempt to leave the moist wound environment.

Do maggots lay eggs inside a wound?
No, only adult flies lay eggs.

Can you feel the maggots crawling on the wound surface?
Many patients do not notice the maggots at all, whereas other can feel them moving about. Full-grown maggots crawling on the skin or edges of a wound can cause tickling or painful sensations. Some patients report pain after the first day or two of maggot therapy. This is probably the result of the maggots—now bigger and heavier—crawling over exposed nerves in the wound bed, or multiple maggots squeezing into small crevices to reach the infected and necrotic debri.

How many maggots should be used?
The number of maggots used depends on the quantity of food available in the wound and the size of the wound. The rule of thumb is to place approximately five to eight larvae per cm^2 of wound surface area.

Can maggots be used under compression bandages?
Yes, provided the maggots still have enough breathing space (use a spacer to prop open the bag).

How long does maggot therapy take?
Treatment usually takes one to three weeks. Treatment is ended when the wound is found to be clean and free of all signs of inflammation. At this stage, the maggots often die automatically due to the deficient supply of nutrients. If an intercurrent infection with *Pseudomonas*, *Proteus*, or *Escherichia coli* develops, the patient should be temporarily switched to an effective antimicrobial wound treatment

Are maggots effective in eliminating multiresistant bacteria (e. g., MRSA)?

Maggots are very effective in eliminating staphylococcal infections, even when the pathogens are multiresistant. The microbial changes that make bacteria resistant to antibiotics do not protect them from the antimicrobial actions of medical maggots and maggot enzymes.

Do maggots in wounds produce unpleasant odors?

Gangrenous tissue (infected, necrotic tissue) smells very bad because the bacteria are releasing foul-smelling gaseous organic molecules into the air. Wounds undergoing maggot therapy often smell bad because the maggots are releasing many of these molecules as they liquify the necrotic tissue. Once the treatment is finished, the debrided wound usually smells much better. In large numbers, maggots themselves can produce ammonia-containing byproducts during protein digestion, which can be detected by a keen nose.

How do the maggots influence the production of wound fluid?

The maggots themselves secrete digestive fluids to dissolve the necrotic wound, and the host responds to the maggots by increasing the wound drainage. In this way, the liquefied necrotic tissue, bacteria, and toxins are flushed from the wound surface and absorbed by the wound dressings (including the Biobag). The increased production of wound fluid is a characteristic sign that the maggots are working.

Is maggot debridement therapy painful?

Most patients do not feel the maggots at all, or else they report that wound pain decreases or disappears after application of the maggots. Some patients report pain after the first day or two of maggot therapy. This is probably the result of the maggots—now bigger and heavier—crawling over exposed nerves in the wound bed, or multiple maggots squeezing into small crevices to reach the infected and necrotic debri. Pain usually can be predicted to occur in patients with significant wound pain before maggot therapy. In some cases, when the wound pain is not adequately relieved by simple pain medication, then maggot therapy must be discontinued early in order to terminate the pain immediately.

Must maggot debridement therapy be performed in a hospital?
No. Outpatient treatment is often preferable, as this eliminates the risk of hospital hygiene problems and lowers the costs of maggot therapy.

Is maggot debridement therapy expensive?
Although the logistics and production standards for rearing disinfected maggots are high, maggot therapy can lower the costs of wound management tremendously. In some cases, only a single application of maggots is needed to effectively debride and stimulate the healing of wounds that have failed to respond to weeks or months of conventional management.

Is maggot therapy covered by my health insurance?
Most health insurance providers and national health plans in Europe reimburse the costs of maggot therapy because it is an effective, low-cost treatment option. In the United States, many but not all insurance plans cover the cost of maggots and maggot therapy. If in doubt, the patient should check with his or her insurance company.

How should maggot dressings be handled at home?
It is important to keep the dressings moist enough to prevent the maggots from drying out; but not so wet that they are saturated or that air can not easily reach the maggots within. Maggot debridement therapy dressings can be moistned with a little sterile water or saline solution, 3 times a day. If the dressing becomes soiled, only the outer secondary gauze dressing should be changed. The inner primary dressing or Biobag containing the maggots should be left in place.

Can I take a shower with the maggot dressing in place?
Patients can shower as long as the maggots are protected from direct contact with water, (e. g., by covering the dressing with plastic wrap). Just remember: maggots can suffocate or drown.

What do I do if the maggots escape from the dressing at home?
A damaged maggot dressing should be removed. If maggots escaping from the dressing are accidentally washed down the drain, they do not create any waste water problems because they are biologically degradable. The damaged dressing should be placed in a plastic bag, securely sealed, and deposited in a normal household waste container.

Suggested Reading

Specialized Literature

Adams GW. Wartime Surgery. In: Doctors in Blue: The Medical History of the Union Army in the Civil War. New York: Henry Schuman; 1952:112–129.

Baer WS. Viable antisepsis in chronic osteomyelitis. Proc Interstate Postgrad Med Assem North Am. 1929;5:365.

Baer W. The treatment of chronic osteomyelitis with the maggot (larva of the blowfly). J Bone Joint Surg. 1931;13:438.

Buchman J, Blair JE. Maggots and their use in the treatment of chronic osteomyelitis. Surg Gynecol Obstet. 1932;55:177–190.

Chernin E. Surgical Maggots. South Med J. 1986;79:1143–1145.

Crosskey RW. Introduction to the Diptera. In: Lane RP, Crosskey RW, eds. Medical Insects and Arachnids. London: Chapman and Hall; 1995.

Cunningham HH. Surgery and infections. In: Doctors in Gray: The Confederate Medical Service. Glouchester, MA: Peter Smith;1970:218–246.

Erdmann GR, Khalil S. Isolation and identification of two antibacterial agents produced by a strain of *proteus mirabilis* isolated from larvae of the screwworm (*Cochliomyia Hominivorax*)(Diptera: Calliphoridae). J Med Entomol. 1986;23(2):208–2.

Fine A, Alexander H. Maggot Therapy—Technique and Clinical Application. J Bone Joint Surg. 1934;16:572–582.

Fleischmann W, Russ M, Moch D, Marquardt C. [Biosurgery—Maggots, are they really the better surgeons?]. Chirurg. 1999;70:1340–1346.

Goldstein HI. Maggots in the treatment of wound and bone infections. J Bone Joint Surg. 1931;13:476–478.

Graninger M, Grassberger M, Galehr E, et al. Biosurgical debridement facilitates healing of chronic skin ulcers. Arch Intern Med. 2002;162:1906–1907.

Grassberger M, Fleischmann W. The BioBag—A new device for the application of medicinal maggots. Dermatol. 2002: 204(4):306.

Grassberger M. A historic review on the therapeutic use of sterile fly larvae. NTM—International Journal of History and Ethics of Natural Sciences, Technology and Medicine. 2002;10:013–024 (German).

Greenberg B. Model for the destruction of bacteria in the midgut of blow fly maggots. J Med Entomol. 1968;5:31–38.

Hoffmann JA, Hetru C. Insect defensins: inducible antibacterial peptides. Immunology Today. 1992;13(10):411–415.

Mumcuoglu KY, Ingber A, Gilead L, et al. (1998) Maggot therapy for the treatment of diabetic foot ulcers. Diabetes Care. 21(11):2030–1.

Mumcuoglu KY, Ingber A, Gilead L, et al. Maggot therapy for the treatment of intractable wounds. Int J Dermatol. 1999;38(8):623–7.

Mumcuoglu KY, Miller J, Mumcuoglu M, Friger M, Tarshis M. Destruction of bacteria in the digestive tract of the maggot of *Phaenicia sericata* (Diptera: Calliphoridae). J Med Entomol. 2001;38(2):161–6.

Mutschler W, Burri C, Plank E. Experimental and clinical experiences with the synthetic skin cover of polyvinylalcohol-formaldehyde foam (PVA). Helv Chir Acta. 1980;47(1–2):163–6.

Pavillard ER, Wright EA. An antibiotic from maggots. Nature. 1957;180(4592):916–917.

Pechter EA, Sherman RA. Maggot therapy: The Medical Metamorphosis. Plast Reconstr Surg. 1983;72(4):567–570.

Prete PE. Growth effects of *Phaenicia sericata* larval extracts on fibroblasts: mechanism for wound healing by maggot therapy. Life Sci. 1997;60(8):505–10.

Robinson W. Stimulation of healing in non-healing wounds by allantoin occurring in maggot secretions and of wide biological distribution. J Bone Joint Surg. 1935;17:267–271.

Robinson W. Ammonium bicarbonate secreted by surgical maggots stimulates healing in purulent wounds. Am J Surg. 1940;47:111–115.

Robinson W, Norwood VH. Destruction of pyogenic bacteria in the alimentary tract of surgical maggots implanted in infected wounds. J Lab Clin Med. 1934;19:581–586.

Sherman RA. A new dressing design for use with maggot therapy. Plast Reconstr Surg. 1997;100(2):451–6.

Sherman RA. Wound myiasis in urban and suburban U.S. Arch Intern Med. 2000;160:2004–2014.

Sherman RA. Maggot therapy for foot and leg wounds. International Journal of Lower Extremity Wounds. 2002;1:135–142.

Sherman RA. Maggot versus conservative debridement therapy for the treatment of pressure ulcers. Wound Repair Regen. 2002;10(4):208–14.

Sherman RA. Maggot therapy for treating diabetic foot ulcers unresponsive to conventional therapy. Diabetes Care. 2003;26(2):446–51.

Sherman RA, Hall MJ, Thomas S. Medicinal maggots: an ancient remedy for some contemporary afflictions. Annu Rev Entomol. 2000;45:55–81.

Sherman RA, Sherman J, Gilead L, Lipo M, Mumcuoglu KY. Maggot debridement therapy in outpatients. Arch Phys Med Rehabil. 2001;82(9):1226–9.

Sherman RA, Wyle FA. Low-cost, low-maintenance rearing of maggots in hospitals, clinics, and schools. Am J Trop Med Hyg. 1996;54(1):38–41.

Sherman RA, Wyle FA, Thrupp L. Effects of seven antibiotics on the growth and development of *Phaenicia sericata* (Diptera: Calliphoridae) larvae. J Med Entomol. 1995;32(5):646–9.

Sherman RA, Wyle F, Vulpe M. Maggot therapy for treating pressure ulcers in spinal cord injury patients. J Spinal Cord Med;1995:18(2):71–4.

Simmons SW. A bactericidal principle in excretions of surgical maggots which destroys important etiological agents of pyogenic infections. J Bacteriology. 1935;30:253–267.

Stevens J, Wall R. The Evoloution of Ectoparasitism in the Genus *Lucilia* (Diptera: Calliphoridae). Int J Parasit. 1997;27(1):51–59.

Teich S, Myers RAM. Maggot therapy for severe skin infections. South Med J. 1986;79:1153–1155.

Thomas S, Andrews A. The effect of hydrogel dressings on maggot development. J Wound Care. 1999;8(2):75–7.

Thomas S, Jones M. Wound debridement: evaluating the costs. Nurs Stand. 2001;15(22):59–61. Review.

Thomas S, Andrews AM, Hay NP, Bourgoise S. The anti-microbial activity of maggot secretions: results of a preliminary study. J Tissue Viability. 1999;9(4):127–32.

Thomas S, Wynn K, Fowler T, Jones M. The effect of containment on the properties of sterile maggots. Br J of Nurs. 2002; 11:S21–S28.

Vistnes LM, Lee R, Ksander GA. Proteolytic activity of blowfly larvae secretions in experimental burns. Surgery. 1981;90:835–841.

Wayman J, Nirojogi V, Walker A, Sowinski A, Walker MA. The cost effectiveness of larval therapy in venous ulcers. J Tissue Viability. 2001;10:91–94.

Wolff H, Hansson C. Larval therapy for a leg ulcer with methicillin-resistant *Staphylococcus aureus*. Acta Derm Venereol. 1999;79:320–335.

Ziffren SE, Heist HE, May SC, Womack NA. The secretion of collagenase by maggots and its implication. Ann Surg. 1953;138:932–934.

Sterile Maggot Suppliers

Polymedics Bio Products
Ambachtslaan 1031
Peer
B-3990 Belgium
Tel.: +32-11-636863
Fax: +32-11-636578
E-mail: info@polymedics.com
http://www.polymedics.com

Surgical Material Testing Laboratory (SMTL)
Princess of Wales Hospital
Coity Road
Bridgend,
South Wales CF31 1RQ
Tel.: +44-1656-752820
Fax: +44-1656-752830
E-mail: maggot-info@smtl.co.uk
http://www.smtl.co.uk/WMPRC/Biosurgery

Neocura
Scientific and Medical Institute
Marktwiesenstr. 55
D-72770 Reutlingen
Germany
Tel.: +49-7071-910388
Fax: +49-7071-910389
http://www.neocura.de

Biomonde GmbH & Co. KG
Sellhopsweg 1
22459 Hamburg
Germany
Tel.: +49-40-55905-342
Fax: +49-40-55905-710
E-mail: biomonde@stathmann.de
http://www.biomonde.de

Further Information Sources

International Biotherapy Society (IBS)
http://www.homestead.com/biotherapy/

Dr. Ronald Sherman
http://www.ucihs.uci.edu/com/pathology/sherman

Traumatology Bietigheim, Germany
http://www.unfallchirurgie-bietigheim.de

Bio Therapeutics, Education, and Research Foundation
http://www.bterfoundation.org

Glossary

adjuvant   assisting or supporting.
adult   fully grown; in this context, it means the actual fly as opposed to the fly larva.
alkalization   modification of the pH of an environment from the acidic range to the basic range.
antibiotic   any one of a class of substances that inhibits the proliferation of or kills micro-organisms, such as bacteria or fungi, without intervening in the biological processes of higher organisms.
antimicrobial   combating micro-organisms.
application   local drug administration.
autoaggression   an immune system attack on the body's own cells.

bacteria   strain of unicellular organisms without a true nucleus.
Biobag   teabag-like maggot containment system constructed using thin sheets of polyvinylalcohol (PVAL) net, which are glued together over sterile larvae and a small cube of a spacer material (PVA sponge).
biosurgery   use of sterile maggots (or substances secreted by them) to debride wounds and combat wound infection.
biotherapy   the use of live organisms in medicine.
blowfly   a genus of flies belonging to the family Calliphoridae; they "blow," or lay their eggs, on decaying flesh.

Calliphoridae   the family of blowflies (shiny metallic flies).
cephalopharyngeal skeleton   hook-shaped structure (mouth hooks) on maggots, used for locomotion and feeding.
causal   related to a cause, directed against a cause.
coli   see *E. coli.*
contaminated   impure.
cuticle   external chitin shell of an insect.

decubitus, decubitus ulcer   see *pressure ulcer.*
dermatological   pertaining to the skin or diseases of the skin.

disinfection   chemical or physical killing of micro-organisms, e. g., through application of alcohol or heat.
diabetes mellitus   syndrome caused by inadequate insulin secretion.
diabetic   pertaining to, caused by, or accompanying diabetes mellitus.
digestive secretions   see *secretion*.
Diptera, dipteran   [*di-pteron* two-winged] the order of true flies within the family of insects.

E. coli   abbreviation for *Escherichia coli*.
enzymes   biocatalysts without which many chemical reactions could not take place at environmental or body temperature.
epithelium   a closed layer of cells that covers the internal or external surfaces of the body ("skin").
Escherichia coli   a ubiquitous intestinal micro-organism and important constituent of the intestinal flora; can also cause disease. Called also *coli bacterium*.
exudate   fluid that escapes during the course of an inflammation.

flora   actually refers to the plant kingdom, but is also used to denote micro-organisms, as in bacterial flora, wound flora, or intestinal flora.
Fournier's gangrene   an infection of the scrotum, penis, and perineum; often has a poor prognosis.

gangrene   death of tissue caused by injury or loss of vascular supply (dry gangrene). If subsequently infected by putrefaction-inducing bacteria, the necrotic tissue is decomposed and liquefied (moist gangrene).
granulation tissue   the thin layer of tissue that generally covers the surface of a normally healing wound within a few days, thereby permitting new skin growth and wound healing.
green blowfly   see *Phaenicia*.

hydrocolloid dressing   a self-adhesive, waterproof wound-dressing material made of gel-forming hydrocolloids with absorbent properties; promotes moist wound healing (autolytic debridement). Used to cover the edges of a wound as part of the primary dressing. The adhesiveness of hydrocolloid dressings supplied by different manufactures can vary.

indication reason or cause for carrying out a given treatment measure after weighing its expected benefits against the potential risks.

L. sericata see *Lucilia* and *Phaenicia*.
labial glands salivary glands that produce digestive secretions and enzymes.
leg ulcer ulceration of the lower leg due to a circulatory disorder (arterial, venous, or both).
Lucilia a genus of flies included in the blowfly family. *Lucilia sericata*: green blowfly, greenbottle; classified as *Phaenicia sericata* in the United States.

malum perforans pedis the presence of deep ulcers of the foot, especially on the heel or ball of the toes; common in diabetes mellitus.
metamorphosis transformation of an organism from its larval form to its adult form, as seen in the maturation of a maggot to an adult fly or of a tadpole to an adult frog.
microbes see micro-organisms.
micro-organisms microscopic animal or plant organisms, including unicellular organisms, bacteria, fungi, and algae.
MRSA methicillin-resistant *Staphylococcus aureus*; antibiotic-resistant micro-organisms encountered in clinical medicine.
myiasis a disease caused by infestation of the body with maggots.

necrosis local tissue death in a living organism resulting from an injury, infection, or other disorder.
necrotic dead

osteomyelitis inflammation of the bone.

5 Appendix

PAOD peripheral arterial occlusive disease; a circulatory order caused by obstruction of a major artery. See also leg ulcer.

parasite an organism that lives upon another plant or animal, drawing its food from the host.

parasite, facultative an organism that is sometimes parasitic, but is basically able to live independent of its host if other sources of food are available.

parasite, obligatory an organism that is always parasitic and cannot survive, grow, or reproduce apart from its host.

pathogen a micro-organism that produces disease.

pathophysiologic pertaining to the dysfunctional metabolism.

penicillin the first antibiotic, discovered in 1928. It is produced by fungi of the genus *Penicillium*, which use the antibiotic to kill bacteria.

Phaenicia a genus of flies included in the blowfly family. *Phaenicia sericata*: green blowfly, greenbottle; classified as *Lucilia sericata* in Europe and many parts of the world other than the United States.

physiological pertaining to the normal, healthy metabolism.

pressure ulcer an ulceration of the skin caused by prolonged pressure, e. g., in a bedridden patient, resulting in local blood flow inadequacy.

primary dressing the inner dressing that holds maggots in the wound.

proteolytic able to decompose or dissolve protein.

Proteus a genus of bacteria that commonly resides in the human intestine; it is often involved in the development of infection and suppuration.

Pseudomonas a ubiquitous genus of bacteria; it is often present in purulent wound infections, where it frequently acts as a multi-resistant pathogen.

remodeling the final reconstruction phase of wound healing; it can last up to two years and makes scar tissue more durable.

saprophage any organism that feeds on carrion.

Sarcophagidae the family of flesh-flies (checkerboard pattern).

secretion   any substance produced by secretion, e. g., wound fluid; in the narrower sense, any product secreted by a gland (e. g., digestive secretions).

secondary dressing   the outer dressing that covers the maggot-containing primary dressing and absorbs the secreted fluids.

sepsis   blood poisoning; a condition that occurs after bacteria spreads from a focus of infection to the bloodstream.

spicule   one of the fine hook-like appendages on the surface of the maggot body.

Staphylococcus   a genus of bacteria, many strains of which are resistant (insensitive) to various antibiotics.

sterile   germ-free, not containing micro-organisms.

stigma   the breathing aperture of maggots.

Streptococcus   a genus of bacteria.

symbiosis   the co-existence of different organisms for their mutual benefit.

synanthropy   the joint development [of an organism] together with humans and domestic animals during the course of evolution.

synergism   the combined action of multiple factors so that their joint effect is greater than the sum of their individual effects.

toxic   generally noxious to living organisms.

ulcer, ulceration   a variable sore accompanied by the loss of skin and tissue substance.

vacuum-assisted wound closure   a sponge dressing system used for local drug administration in wound management.

vacuum therapy   the application of negative pressure to a wound surface using a vacuum device with a sponge dressing; it is used to stimulate wound healing.

Index

A

adverse effects, maggot therapy 63–64
American Civil War 16–17
amputation, and maggot therapy 51, 53–54
 in diabetic patients 37
analgesia 44
anatomy, maggot **10**, **13**
antibiotics 49, 59
antimicrobial action 24–25, 26
arterial occlusive disease 27
auto-disinfection 13

B

Baer, W.S. 17–20, 58
bedsores 45–48
Biobags 35–36, 59
 and chronic wound infection 55
 and decubitus ulcer **47**
 and foot wounds **41**
 and future of maggot therapy 65
 hysterectomy wound treatment 57, **58**
 and leg ulcer **44**
 outpatient handling 69
 preparation **35**
 schematic view **36**
blowfly, term explained 4, 8
 see also flies; Phaenicia sericata
bone infection 54
breathing, maggots 9, 10

C

Calliphoridae 8, 9
case reports
 diabetic gangrene 38–41
 leg ulcer 43–45
 pressure ulcers 47, 48
 wound infection 50–52, 55–58, 59–63
Charcot's syndrome 38
clinical application, maggot therapy 32–65
compression bandages 67
containment bag 34–37, 46, 65
costs, maggot debridement therapy 65, 69

D

DeBakey, Michael 53
debridement 23, 26, 29
 see also maggot debridement therapy
decubitus ulcer 45–48, **47**
diabetic gangrene **39**, **40**
diabetic wounds 37–41
digestive enzymes, of maggots 11
disease transmission, flies 7–8
disinfection 24
dressings, maggot
 containment bag 34–37
 discarding 34
 free-range system 32–34, **33**, 51
 outpatient handling 69
 primary 33
 secondary 34
 used in 1934 **19**
 waste disposal 34, 64, 70

E

efficacy, maggot therapy 64
eggs 4–6
 hatching 6
 and maggots 67
 Phaenicia sericata **5**
energy stores, in maggot 13

F

feeding and growth, in maggots 10–12
femur infection 59–63, **60, 61, 62**
Fleming, Alexander 21
flies 3–4
  blowfly, term explained 4, 8
  disease transmission 7–8
  as fomites 7
  life cycle 4–7
  as vectors 7
  see also larvae; maggots; pupation
fomites 7
food storage, in maggots 11
Fournier's gangrene 27–28

G

gangrene 27–28, 52
  diabetic 38–41
granulation tissue 25, 57
green blowfly, *see Phaenicia sericata*
greenbottle flies 8
growth of maggots 10–12

H

Hippocrates 2
hirudin 1
history of maggot therapy 14–16
  American Civil War 16–17
  World War I 17–21
Hortus Sanitatis (1491) **14**
hospitals, and maggot containment 63–64
hysterectomy, wound infection 55, **56, 57, 58**

I

immune system 52
infection 63–64, 66
instar larvae **5, 32**
International Biotherapy Society 22
intestinal tract, of maggot 12–13

J

Jones, J. 17

K

Keen, W.W. 16
Koch, R. 17

L

Larrey, D.-J. 15, 16
larvae, *Phaenicia sericata* **5,** 9–13, **11, 12, 32**
Lederle Laboratories **20,** 21
leg ulcers 38–45, **44, 45**
life cycle, maggots 6, **7**
light, and maggot movement 9
Linnaeus 3
Lucilia, see Phaenicia

M

maggot debridement therapy 2, 22, 23
  adverse effects 63–64
  clinical studies 21–22, 65
  with contained maggots 34–36
  costs 65, 69
  and diabetic gangrene 38–41
  dressings, *see* dressings
  future 64–65
  history 14–16
  and hysterectomy wound **58**
  indications 37
  and leg ulcers 42
  mechanisms 26
  and osteomyelitis 63
  outpatient 60
  and pain 44, 63
  and pressure ulcers 46
  psychological complications 64
  revival 21–22
  risks 63–64, 66
  safety 59
  treatment period 67
maggots 3–13

anatomy 9, **10**, **13**
antimicrobial action 24–25
breathing 9, 10
clinical applications 32–65
digestive enzymes 11
energy stores 13
escape 34, 63–64, 69, 70
feeding 10
head **10**
and infection 66
intestinal tract of 12–13
life cycle 6, **7**
medicinal uses 27–28
modes of action 22
refrigerating 33–34
spicules 23, **23**
sterile 21, 28–29, **29,** 75
and wound healing 14–31
metamorphosis 4
methicillin-resistant *Staphylococcus aureus* (MRSA) 26, 49 50, 68
military medicine, and maggot therapy 14–21
mode of action, maggot therapy 26
myiasis 29–31

O
odor, wounds 48, 56, 68
Old Testament 3–4, 14
osteitis **60**
osteomyelitis 63
outpatient treatment 69
ovipositor 4, **6**

P
pain, and maggot therapy 44, 63, 68–69
parasites, and flies as vectors 7
Paré, Ambroise 4, 15, **15**
Pasteur, Louis 17
penicillin 21
Phaenicia sericata **8,** 8–9
adult female **6**

eggs **5**
feeding and growth 10–12
larvae **5,** 9–13, **11, 12, 32**
sterile, production 28–29
see also flies; maggots
pharmaceutical industry 1
phototaxis, negative 9
pressure ulcers 45–48, **47**
Proteus spp. 25, 26, 43, 63
Pseudomonas spp. 43, 63
pupation 6, **11**

R
refrigeration, of maggots 33–34
risks of maggot therapy 63–64, 66

S
showering, with maggot dressing 69
skin irritation 63
soft-tissue infections 58
spicules 23, **23**
spiracles 9
Staphylococcus spp. 53, 58
methicillin-resistant 26, 49, 50, 68
sterile maggots 21, 28–29, **29,** 75
Streptococcus spp. 58
Surgical Maggots-Lederle (1930) **20**
synanthropy 3

T
tetanus 52
thigh infection 59–63, **60, 61, 62**

V
vacuum-assisted wound closure 49
vacuum therapy 45

W
waste, dressing disposal 34, 64, 70
World War I 17–21
wound healing

and maggots 14–31
  stimulation 25–27, **26**
wound infection
  acute 49–54, **50, 51, 52, 53**
  chronic 54–63
  hysterectomy 55, **56, 57, 58**
wounds
  debridement, *see* debridement;
  maggot debridement therapy
  diabetic 37–41
  disinfection 24
  fluid production and maggots 68
  future of maggot therapy 64
  maggot-infested 30–31

Z

Zacharias, J.F. 16
zinc paste 32